TAXING AIR

FACTS & FALLACIES ABOUT CLIMATE CHANGE

TAXING AIR

FACTS & FALLACIES ABOUT CLIMATE CHANGE

Bob Carter & John Spooner

with

Bill Kininmonth

Martin Feil

Stewart Franks

Bryan Leyland

kelpie
press

Windmills are an expensive and wasterful response to an improbable threat. June 2007.

DID YOU KNOW THAT ...

Just 8,000 years ago, there was virtually no sea-ice in the Arctic Ocean

More polar bears live now than at any time since surveys began

Sea-level rise is natural, and declining in rate

Australian rainfall has not decreased over the last 100 years

A previous Australian drought lasted 69 years

The Murray-Darling Basin now contains 3 times as much
water as it held in its natural state

Global air temperature has not increased for the last 16 years,
despite a nearly 10% increase in CO_2

Global ocean temperature is also steady or cooling slightly

As a nation, our territory absorbs up to 20 times
the amount of CO_2 that we emit

The federal government does not audit the scientific advice
that it receives on climate change

The CO_2 tax will cost you about $1,000 a year — and rising

The result of reducing Australian CO_2 emissions by 5%
by 2020 will be a theoretical (and unmeasurable) cooling
of between 0.0007° and 0.00007° C

No scientist can tell you whether the world will be
warmer or cooler than today in 2020

Praise for *Taxing Air*

Taxing Air is an outstanding contribution to the growing literature that examines and calls to account the climate alarmism of the past two decades. Written for the lay person and aided by Spooner's insightful illustrations, it provides an accurate, easily understood explanation of the many scientific and technical issues that comprise today's climate science. Equally important, it examines the history and exposes the duplicity of some of the individuals and organisations who have vested interests in creating and maintaining horrific visions of an imagined global warming future. The book would make a splendid gift to certain members of the climate science establishment.

Dr. Art Raiche,
CSIRO chief research scientist (retired)

After starting to read it, I could not put Taxing Air *down. My congratulations to the team of authors, who have highlighted every facet of the worldwide scam that is called Man Made Warming, and which demands that all families pay a tax on the very air they breathe. Responding to vested interests, western politicians have already wasted trillions of dollars to frighten people with lies about industrial carbon dioxide emissions.*

In fact, today's global temperature lies well within life's limits – indeed, the present-day is cooler than much of previous geological time. The gas of life rather than a pollutant, atmospheric carbon dioxide has nurtured all the forms of organism on planet Earth for many hundreds of millions of years, as is so clearly explained in this beautifully written and illustrated book.

Professor David Bellamy, OBE
President, Conservation Foundation, UK
Trustee, World Land Trust (1992-2002)

Published in Australia by Kelpie Press
glrmc42@gmail.com

National Library of Australia
Cataloguing-in-Publication entry

Author: Carter, Robert; Spooner, John.
Title: Taxing air: facts and fallacies about climate change /
Other authors/contributors: Feil, Martin, author; Franks, Stewart W., author;
Kininmonth, William, author; Leyland, Bryan, author;

ISBN: 9780646902180 (paperback)

Notes: Includes bibliographical references and index.
Subjects: Global warming--Australia.
  Climatic changes--Australia.
  Climatic changes--Environmental aspects--Australia.
  Greenhouse effect, Atmospheric--Australia.
  Carbon taxes--Law and legislation--Australia.

551.6

All cartoons were originally published in *The Age*, Melbourne.
Figures designed by Frank Maiorana, FM Visuals, Melbourne.
Designed and produced by High Horse Books, *www.highhorse.com.au*

For the last 20 years, the received scientific wisdom about global warming has been provided to governments by the Intergovernmental Panel on Climate Change (IPCC).

This book is dedicated to the army of independent persons and bloggers who — by questioning every precept, analysing every extravagant claim and insisting on the importance of empirical evidence — have helped to keep the IPCC honest and the spirit of true scientific enquiry alive.

History will salute them.

Contents

settled? • But don't 97% of all scientists say that dangerous warming is occurring? • But isn't there meant to be a consensus about global warming? • Is there any common ground amongst scientists who argue about this matter? • What was the Kyoto Protocol?

How do we know about ancient climate? • What is a proxy record of temperature? • How do we measure modern temperatures? • How long is the record of worldwide direct measurements of temperature? • Over what time periods does temperature change reflect climate change? • What are Milankovitch variations? • Does melting ice mean that global warming must be occurring? • What about other circumstantial evidence: coral bleaching or polar bears, anyone? • What is the Holocene and why is it important? • What were the Medieval Warm Period and Little Ice Age?

Is the Earth in climatic equilibrium? • What is a greenhouse gas? • The Sun obviously warms the Earth, but what cools it? • Is carbon dioxide the most important greenhouse gas? • What is the classical explanation for the greenhouse effect? • How is the greenhouse effect now understood by scientists? • What is the greenhouse effect as understood by the general public? • Is less warming bang really generated by every extra carbon dioxide buck? • What is climate sensitivity? • What net warming will be produced by a doubling of carbon dioxide? • Do changes in carbon dioxide precede or follow temperature change? • How can the hypothesis of dangerous greenhouse warming be tested? • Is atmospheric carbon dioxide a pollutant? • Are modern carbon dioxide levels unusually high, or dangerous? • What about methane, then? • Do I have to worry about ozone too?

What is a climate model? • What are the main types of climate model? • Why do we need computer models to study climate change? • But can computer models really predict future climate? • Do computer models suggest 'fingerprints' for human-caused warming?

Can we take the temperature of the ocean? • Why is it important to distinguish between local and global sea-level change? • What controls the position of the shoreline at Bondi Beach? • Is it true that Australia is going to be swamped by rising sea-levels? • What part does the ocean play in controlling climate? • Is there such a thing as ocean acidification, and should we worry about it?

emissions? • Why is Labor so certain that its carbon dioxide tax can't be repealed? • So can the Coalition keep its promise to unwind the tax? • But aren't we doing all this for our children and grandchildren?

Electrons are clean: what on earth is all this about dirty energy? • How much use is windpower? • Why aren't windfarms a cost-effective source of base-load electricity? • But surely windfarms are environmentally beneficial? • Well, at least windfarms save carbon dioxide emissions, don't they? • What about solar power, then? • Perhaps tidal power is the solution? • How do biofuels benefit the environment? • Why isn't nuclear energy part of Australian and NZ energy planning? • What's WRONG with coal-fired power stations, anyway?

Surely we should give the Earth 'the benefit of the doubt' about global warming? • What has climate change got to do with energy supply anyway? • What are Australia's greatest natural hazards? • What does the climatic future really hold? • Do we really need a national climate policy, then? • What can I do to help achieve a sensible national climate policy?

Preface

It has been estimated that more than 100 areas of knowledge are relevant to discussions of human-caused global warming, and of the related, broader issue of climate change in general. Many but not all of these specialities are scientific in nature.

An inevitable consequence of this is that there is no such thing as an overall expert on climate change. Instead, many qualified people from different backgrounds are experts, more or less, on small and generally different parts of the issue.

It is against this reality that we decided to write this book, the six contributors to which have, of course, varied backgrounds and varied qualifications (p.255). We are respectively a journalist and cartoonist (Spooner), a geologist and palaeoclimatologist (Carter), an economist (Feil), a hydrological engineer (Franks), a meteorologist-climatologist (Kininmonth) and a power-systems engineer (Leyland), and our life experiences have included employment in the newspaper, government department, academic and business consultancy areas.

Because the book is a genuinely collaborative effort, we have chosen not to allocate particular answers to questions, or sections, to individual authorship. Rather, the four scientists and engineers among us take collective responsibility for the scientific and engineering statements made in the book, Martin Feil has mostly written and vouches for the material that relates to economic and social issues, and John Spooner, in addition to the introductory essay, cartoons and illustrations, has provided constant enthusiasm and good-natured editorial urging whenever it was needed.

To ensure maximum readability we have deliberately chosen not to clutter the text with references to the individual publications that support the statements of fact that we make, though where essential brief technical explanations have been provided in footnotes. Readers interested in greater scientific detail, and supporting literature, should consult the Recommended Reading afternote (p.251), which supplies some carefully chosen and comprehensive sources of more detailed information.

Given the emotionalism that accompanies much of the public debate about global warming, it is important that we state that none of the authors has any financial vested interest in a particular policy outcome from the climate debate. Specifically, too, none of us has been paid by vested interests to express particular views on global warming, be that in this book or elsewhere.

We do, however, share one strong vested interest, which is in the provision of balanced, academically rigorous and accurate analysis of important public policy issues that are science-related. This has been our aim for the topics of climate change and global warming that are covered in this book.

The publication of any book like this one depends upon the efforts and support of many people other than the authors themselves. First and foremost, we thank the members of our families for their unwavering support, and also our many professional colleagues who, often in ways that they may be unaware of, have helped to guide our thinking – sometimes gently, and sometimes less so! More specifically, we thank Chris de Freitas, Frank Maiorana, John McLean, Jennifer Marohasy, Russ Radcliffe, Ingrida Rocis, Willie Soon and Anne Verngreen for having made contributions to the writing and publishing of this book without which the venture could not have been completed.

The Authors
April, 2013

How the cartoonist got his ideas

A state of scepticism and suspense may amuse a few inquisitive minds. But the practice of superstition is so congenial to the multitude, that if they are forcibly awakened, they still regret the loss of their pleasing vision. Their love of the marvellous and supernatural, their curiosity with regard to future events, and their strong propensity to extend their hopes and fears beyond the limits of the visible world, were the principal causes which favoured the establishment of Polytheism.
Edward Gibbon, *The Decline and Fall of the Roman Empire*

I WAS ONCE approached by a friend who is concerned about the danger of human-caused global warming. He asserted that when it comes to scientific issues of major public concern like this, it is 'not what you believe but who you believe'. I think he meant that my then hesitant scepticism about global warming was pointless, for as a cartoonist I must be as inadequate to judge the science as he was. For that matter it seems all of us who are untrained in 'climate science' have no option but to respect the peer reviewed authority of the climate science establishment. Of course, as a revered public intellectual, my friend did not see it as his duty to sit on his hands. He felt bound, as many have, to vigorously support the scientific and political authority of the Intergovernmental Panel on Climate Change (IPCC), and regional associates like the CSIRO.

I found my friend's advice baffling. As anyone familiar with the judicial process knows, the gravest issues of liberty and fortune are often determined by a jury selected from the general public. During some trials, expert witnesses give evidence supporting either side in our adversarial system. The judge must rule which evidence is relevant or admissible, but in the end it is the jury that decides which version of the scientific evidence is to be believed. No one in a civilised society is daunted by this process. We accept the outcome unless a procedural mistake has been made. Often someone goes to jail because one cross-examined scientific expert is believed over another by ordinary jury persons. No big deal.

So what's the problem? Everyone has the chance to do some reading until they hit the wall of their own ignorance or understanding. Then you ask for help. Acting as the foreman of your own jury, you can ask for more direction or for clarifications to help you follow the logic of the argument. But if in the end you cannot agree with your fellow jurors, then you cannot reach a verdict. It is surely the duty of scientists who wish to influence political events to explain themselves clearly. If they can't do that to the degree that ordinary people (not to mention many of their equally qualified peers) understand and accept that there really is a dangerous global warming problem, then it is premature for governments to be setting expensive anti-carbon dioxide measures in place.

But in matters to do with climate change there is no judge except the scientific method, i.e. the proposition of a testable hypothesis followed by its testing against factual or experimental challenge. That it fails various empirical tests is, of course, precisely why the advocates of dangerous anthropogenic global warming (AGW) are attracted to the idea of a scientific consensus. And that is where things get difficult for cartoonists, public intellectuals, journalists, politicians, bloggers and the general public. The reason why the phrase 'scientific consensus' emerges in this debate is because political activists want to get moving, and if they say that the so called 'scientific consensus' is scary and urgent, then cartoonists and others had better just get out of the way: the science is settled and procrastination is outrageously reckless. The question of whether there is, or can be, such a thing as a useful scientific consensus about a matter like dangerous AGW is a difficult theoretical and practical problem. Cutting through that uncertainty, AGW activists have preferred to use the political process to impose their consensus argument mainly through the media.

The activist cause perhaps peaked in early 2007 when Al Gore's film *An Inconvenient Truth* became an international hit, winning two Academy Awards. This evidence might have seemed compelling to the uninitiated, but in October 2007 the British High Court found the film contained at least nine significant errors of fact and required British schools to refer to these errors when using the film in lessons. Though Professor Bob Carter gave evidence

Al Gore, chairman of the Alliance for Climate Protection and former US vice-president. July 2007

The Great Global Warming Swindle presented a sceptical view of climate change. July 2007.

in this case, to date few people in Australia are aware of this severe embarrassment for Mr Gore. Thanks mainly to media neglect, I certainly never heard of the case at the time, and so, like nearly everyone else, I was initially taken in by the authoritative pronouncements of the former vice-president. However, I remember too that later in 2007, when the ABC broadcast Martin Durkin's provocative documentary, *The Great Global Warming Swindle*, a lot of people got very upset indeed. How interesting. The science was settled; the debate was over; no more discussion was needed, yet all it took was one contrarian TV program to cause an explosive and long-running public sensation.

Any media professional should have been aroused by such an excited censorship campaign, and it stimulated my first cartoon on the subject, which depicted the family television set as medieval stocks with an imprisoned climate sceptic being pelted by the family with their TV dinner.

And what of Durkin's documentary? I know it didn't get one or two of the fine details of the science exactly right, but then very few documentaries ever do. For example, there has been much criticism that any influence of cosmic rays on clouds will apply only to lower level clouds — not all clouds, as the program stated. But, as in many good documentaries, Swindle presented some riveting interviews with high calibre professional scientists. To take one example, we heard from Professor Paul Reiter, chief of the Insects and

Infectious Diseases Unit at the Pasteur Institute in Paris. As if I were in a jury, I had the opportunity to see him as he spoke (remember that Appeals Courts won't hear an appeal based on a written transcript), and I formed a strong impression that he was telling the truth — that mosquitoes are equally at home in freezing Siberia as they are in the tropics. The same goes for malaria, the disease that they carry.

Professor Reiter also seemed credible when he spoke of his difficulties with the IPCC process, describing why he thought that the organisation was dysfunctional. Many of the other interviews with sceptics, including one with the co-founder of Greenpeace, Patrick Moore, had a similar effect on me. Though I hadn't seen all the evidence, after this film was broadcast the sceptics at the very least had my attention, and no doubt that of many other independent persons. So why did the media at large attack the film, and continue to vigorously promulgate their belief in dangerous AGW?

Up to the time of Swindle's screening, the role of journalists in the global warming debate had perhaps been unexceptional. But things changed after the screening of the documentary and the outpouring of protest and criticism that it attracted. Then, and just as lobbyists do for matters of economic or social reform, the proselytisers for global warming alarm, who were feeling threatened as never before, got nasty. Someone came up with the brilliantly clever but insidious idea of using the term 'denier' to describe a person who remained agnostic or sceptical about the extent of human contribution to the global warming of the last 100 years. Why 'denier'? Because it made the connection in people's minds to 'Holocaust denial'. Unbelievably, this malicious rhetoric henceforward came to be adopted by climate activists, media reporters and politicians up to the level of heads of state, and was applied to distinguished science professors such as Paul Reiter, Richard Lindzen, Freeman Dyson, William Happer and many others.

Holocaust denial describes the heartless and despicable refusal by anti-Semites to acknowledge the historical truth of the Jewish genocide that occurred during World War II. If you use the offensive term 'denier', you do so for reasons best known to yourself. You may be calculating or you may be indifferent, but as politicians like Kevin Rudd, Penny Wong and Julia Gillard (all users of the term) would have known, the effect is pungent. No sensible, morally responsible person wants to be stigmatised in such a way.

Intimidation comes in many forms, and there can be no doubt that many people have been inhibited from entering the public debate on dangerous AGW because of the intimidatory power of this vicious language. And just in case you still haven't got it, some prominent Australian public intellectuals to

CITIZENS WONG AND RUDD DENOUNCE A LEADING CLIMATE WARMING DENIER

Prime Minister Kevin Rudd and Minister for Climate Change Penny Wong
never understood the risk of conflating climate change with weather. July 2008.

this day continue to explicitly endorse the moral equivalence between Holocaust and global warming denial. This endorsement is all the more incredible
because it comes from academics who really understand the horror of the
Holocaust.

Nonetheless, this blatant method of stigmatising those who questioned
the so-called 'consensus' view on AGW turned out not to be enough to suppress all independent views: many agnostic and sceptical scientists are made of
tougher stuff. Accordingly, more stops had to be pulled out on the vilification
organ, with sceptical scientists being compared by journalists and Labor politicians to 18th century slave trade advocates, the odious tobacco lobby and
recently even to the stench of paedophilia — a new low in public discourse.

Every cartoonist and satirist in the world, not to mention the investigative reporters, should by now have had their bullshit detectors on high alert.
If the evidence was so good, and the sceptical scientists were so weak, wrong
and few in number, then why the need for such rancorous politics? If you
have the UN, the EU, the banks, the financial markets, most of the clergy and
the media on your side, then why this dishonourable nastiness as well? I've

always hated bullies and they have certainly been thick on the ground in this debate.

No good came of going back to my friend, the intellectual, for another discussion. I was referred to the 'only' source of definitive knowledge on the subject; which was supposed to be Dr James Hansen, head of the NASA Goddard Institute for Space Studies. By this time, the alarm bell in my head was ringing loudly. For example, I came to know that only 0.7°C of warming had occurred since 1910, and only 0.4°C of that since 1945 when carbon dioxide levels started to rise considerably. Next, I was told that the Arctic sea ice was melting in an 'unprecedented' way, despite abundant scientific documentation that the Arctic Ocean was virtually ice free during the Holocene climatic optimum, only eight thousand years ago — nonetheless, the public continued to be told that this melting sea ice manifests a 'tipping point' that will lead to catastrophe.

These things notwithstanding, if NASA was the 'main authority' then I thought that I should do as I was advised and consult them. So I went to the NASA web site and searched for material on Arctic sea ice melt. Five items into that page I discovered that 'a new NASA-led study' into the causes of Arctic sea ice melting had reported 'a 23% loss' in the Arctic's year-round sea ice cover between 2005 and 2007. The research team was led by Dr Son Nghiem of NASA's Jet Propulsion Laboratory in California, who said that the rapid decline in winter perennial ice over the previous two years had been caused by unusual winds, which 'compressed the sea ice … and then sped its flow out of the Arctic' where it rapidly melted in warmer waters. Dr Nghiem also said that 'the winds causing this trend in ice reduction were set up by an unusual pattern of atmospheric pressure that began at the beginning of this century'. Yet nowhere in the public discussion of the dramatic sea ice melt had we heard much about these real causes for its diminution, including the North Atlantic Oscillation (NAO), doubtless because it was easier to concentrate on the positive feedback loops created by the exposed Arctic sea. The real story was far more complicated and difficult to explain than in the glib terminology of AGW.

Furthermore, other than on Andrew Bolt's and Joanne Nova's blogs, it is extremely hard to find any widely read popular public description in Australia (or worldwide for that matter) of the melting of Arctic sea ice that occurred between 1920 and 1940. Virtually all media coverage has related to the short satellite record of Arctic sea ice, which is only available since 1979. Rather than a simplistic alarming story of global warming the NASA research was an accurate account of only part of a complex matter.

The real story in the Arctic Ocean obviously involves an intricate relationship between sea ice, ocean currents, atmospheric winds and temperature as affected by ocean-atmosphere oscillations like the North Atlantic Oscillation, and these modern changes need to be studied in the context of changes that have occurred through millennia. When viewed in these wider contexts, there is nothing untoward about the relatively minor changes in sea ice cover that have occurred in the Arctic Ocean in modern times.

This was, of course, a great time for cartooning. John Howard saw the votes to be gained in crossing over to the warming camp, and Kevin Rudd promised to save the planet from the greatest moral, economic and spiritual threat of our time.

As the Arctic ice melted in 2007, Australia was suffering the continuation of an allegedly unprecedented, decade-long drought. The absence, indeed end, of sufficient rain to fill our rivers and dams was predicted by grim-faced climate scientists who invariably announced that things were far worse than their computer models had predicted. Yet the preceding severe droughts of the 1860s, the 1890s and during World War I were rarely discussed in order to provide a needed perspective. With a simple Google search, anyone could, and still can, access photographs of horses and carriages on a bone dry Murray River bed in 1914.

With so much political clout behind the dangerous warming cause, and the Australian drought in full force, the next game changing moment that captured my attention occurred on December 19, 2007. Dr David White-house caused a stir by writing an article for the left wing *New Statesman* magazine entitled 'Has Global Warming Stopped?' Dr Whitehouse stressed a point which concerned sceptics had long noticed: 'The fact is that the global temperature of 2007 is statistically the same as 2006 as well as every year since 2001.' Because of the fundamental mechanism of global warming (the greenhouse effect), temperatures should have been increasing as carbon dioxide levels continued their relentless rise; but they were not. As Whitehouse, a PhD in Astrophysics and former online science editor for the BBC, noted, 'something else is happening [to the climate] and it is vital we find out what or else we may spend hundreds of billions of pounds needlessly'.

It was about this time that slowcoach denier cartoonists like me really started to wake up and look around, to discover the writings of experienced agnostic scientists like William Kininmonth, a former head of Australia's National Climate Centre at the Bureau of Meteorology. Senior scientists like William had been publishing serious critiques of dangerous AGW way back in the 1990s, which was long before I and other slowcoaches had stopped our dreaming.

A CONVENIENT TRUTH

Prime Minister John Howard ponders the electoral implications of climate change. October 2006.

So just when the those supporting climate alarm thought that they had everything settled and nailed down, a gale of discontent started to blow. Cartoonist heaven, really. We love the spectacle of powerful people preparing their policy vessel against strong winds and rough seas, frantically rigging up fragile, flapping sails of spin and blather. If you're going to spend over $15 billion of taxpayer's money on desalinated water, or manage a potentially ruinous carbon dioxide trading scheme (please don't stock our superannuation with the stuff), then you certainly don't want to be questioned too closely, let alone lampooned, about the scientific details that you misunderstood or got wrong.

Nobody anticipated the next debacle: Climategate. As the result of an apparent hacking attack on a server at the Climatic Research Unit (CRU) at the University of East Anglia (which might, or might not, have been mounted by an internal whistleblower), thousands of emails previously exchanged between senior IPCC scientists were leaked to the public a few weeks before the 2009 Copenhagen Climate Change Conference. The emails reeked of scientific uncertainty, political manoeuvring, unreasonable secrecy and strange ethics. The revelations that they contained undoubtedly exercised an influence on the failure of the Copenhagen conference.

Prime Minister Kevin Rudd, November 2009.

Climategate also prompted at least a short burst of candour from Professor Phil Jones, Director of the CRU, who confirmed in a BBC interview that the warming rates of the periods 1860–1880, 1910–1940, 1975–1998 had been statistically similar; that from 1995 to 2009 there had been no statistically significant global warming; and that from 2002 to 2010 there had been slight but 'insignificant' global cooling. In answer to a further question as to whether the 'climate change (debate) is over', Professor Jones stated, 'I don't believe the vast majority of climate scientists believe this'. I found this statement extremely encouraging, for the science was obviously not settled and the consensus was crumbling even amongst the warming devotees.

Since 2007 the non-scientific players in this great intellectual drama have been confronted by a creeping uncertainty (which some still do not want to acknowledge) concerning many contentious dangerous AGW issues. These have included: the composition of the IPCC and the credibility of its processes; the unusual melting, or not, of sea ice and glaciers; the evidence for medieval warm temperatures; the importance of sunspots; the measurement of claimed global warming; changes or not in patterns of extreme weather events; ocean 'acidification'; ocean warming and sea-level rise; biomass absorption and the longevity of molecules of atmospheric carbon dioxide; the reliability of climate computer models; the influence of the short-period

El Niño Southern Oscillation (ENSO), and other similar oscillations on a multi-decadal scale; the chaotic behaviour of clouds; the impact of cosmic rays on climate; realisation that it is just clean air that is being vented by the Yallourn power stations (carbon dioxide and water, with virtually no pollutants); and, to cap it all off, even a newly declared scepticism towards dangerous AGW by green gurus like James Lovelock, the founder of the Gaia movement.

By early 2010, it seemed that nearly every single element of the global warming debate was well and truly up for grabs.

In addition, and not put off by having their sanity questioned, myriads of qualified agnostic and sceptical persons made public statements, or signed declarations or petitions, to the effect that whilst dangerous AGW was a theoretically possible outcome of human-related carbon dioxide emissions, it was a very unlikely one given that, despite strenuous efforts, no proven AGW at all had yet been identified at a measurable level. For example, in the Oregon Petition, starting in 1998, more than 31,000 scientists, including 9,029 with PhDs, signed a statement of protest at the findings and recommendations of the IPCC.

The IPCC was embarassed when it was revealed their prediction that Himalayan glaciers could melt by 2035 was based on non-scientific evidence. January 2010.

June, 2011

In Australia, and against this hurricane of uncertainty, the tattered vessel of government climate policy heedlessly weighed anchor and began to implement the demonisation of carbon dioxide by introducing penal taxes against its emission. Instead of waiting out the storm in harbour, government activists set out to sea guided by the Green faith and a few bearings taken on scattered windmills along the shoreline.

All of this provided great material for a satirist, but it was very bad news indeed for the average Australian citizen whose cost of living was inexorably on the rise. In addition to the continuing increases in direct costs, it is also painful to contemplate the things that could have been done to improve our schools or health service using the money that has instead been squandered in vain pursuit of irrational renewable energy targets and 'stopping global warming'.

Imagine if the sceptics are right.

Who is going to be accountable, and who is going to do the accounting?

What of the Establishment activists, and their media supporters, who have so vilified a group of honest, brave and experienced scientists for merely staying true to the empirical values of their profession? Who will vindicate the sullied reputations of, to name but a few antipodean names: Michael Asten, Bob Carter, Chris de Freitas, David Evans, Stewart Franks, William Kininmonth, Bryan Leyland, Jennifer Marohasy, John McLean, Joanne Nova, Garth Paltridge, Ian Plimer, Peter Ridd and Walter Starck? And the same question applies also for economists like Henry Ergas, Martin Feil, David Murray and others, who have dared to suggest that the Stern and Garnaut reviews were a travesty on both scientific and economic grounds, and that the carbon dioxide pricing/taxing emperor actually has no clothes.

I would love to see a list of all those socially beneficial environmental, educational and health projects that could have been funded instead of the profligate and futile spending on dangerous AGW that has actually occurred. I would like, too, though I doubt that I will see it in my lifetime, to see a public apology from all those advocates, intellectuals and politicians who have so freely slandered and injured the moral reputations of those other Australian citizens and qualified scientists whom they call 'deniers'.

History is usually written by the political victors, but the global warming issue seems set to continue as a ritualised tribal debate for a long time yet. I once asked a seriously committed 'warming' journalist how many years the present pause in warming would have to last to cause her to challenge her own belief. Calmly looking me dead in the eye, she said 'fifty years'.

John Spooner
April, 2013

1

Climate and Climate Change

What is climate?

Climate is the long term average of the weather, and it shapes our lives.

We all live with the weather every day. Thus we understand intuitively what weather is, and that it is above all changeable. Climate is simply the annually recurring patterns of weather, averaged over the longer term.

Weather and climate manifest themselves as the changing physical factors in the environment that regulate our everyday life. The daily and annual cycles of temperature, the seasonal patterns of rainfall, and the characteristic changing patterns of sunshine, cloudiness and wind all contribute over time to defining the local climate that each of us inhabits. It is the variable nature of local climates over different parts of our planet that determines their characteristic assemblies of microbial, plant and animal life.

The instrumental temperature record of the last 150 years (Fig. 1, p.17) encapsulates subtle changes in weather and seasonal climate patterns but, being short, is not necessarily a harbinger of longer term trends. The observed changes are influenced by a large range of factors, some recognised but many remaining unknown.

Over much longer periods of time, hundreds of thousands to millions of years (Fig. 2, p.17), the drifting passage of continental plates has opened and closed major ocean connections, and thus modified the transport of heat from the tropics to polar regions. In addition, the changing locations and physiography of land masses, accompanied by mountain building and erosion, have from time to time altered the patterns of heat transport in the atmosphere.

The second major influence on long term climatic patterns is that provided by small changes in the geometry of Earth's orbit around the Sun, especially changes in the tilt and precession of Earth's axis relative to the plane of orbit and its eccentricity (see III: What are Milankovitch variations?). The changing tilt causes the dominant 41,000 year-long, cyclic pattern that is apparent in Fig. 2. Of course, the tilt of the Earth's axis is also what causes the seasons of the year, and the strength of seasonality at any one time depends upon the angle of tilt (varying between 22.1° and 24.5°).

The changing eccentricity, as the Earth moves between near circular to elliptical orbits on periods of about 100,000 years, is not a new phenomenon, but it has only exercised a strong impact on climate over the last half-million years or so of enhanced glacial cycles (Fig. 2, p.17). The reasons for this are much discussed and not fully understood.

Other important factors that help to determine climate variability include changes in the radiation, magnetic field and high energy particle streams of the sun and the changing influx of cosmic rays from deep galactic space (see VII: How important are cosmic rays in affecting global climate? What about the sun?)

The key point is that, like weather, climate is always changing. Change is simply what climate does.

What is a climate scientist?
With so many fields of knowledge involved, no one can be an overall 'climate expert'.

When media outlets cover a news story about global warming, they often interview scientists who are said to be 'climate experts'. Yet the complexity of the climate system is so great, encompassing as it probably does many dozens of subdisciplines and exercising an influence on all life on our planet, that no such person as a 'climate expert' actually exists.

Scientists who are interviewed on the media are typically expert in one

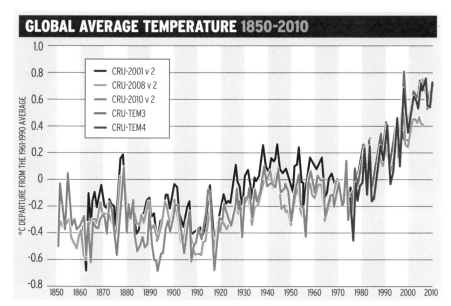

Fig. 1. Global average atmospheric temperature, 1850–2010, compiled from ground thermometer measurements (U.K., Hadley Centre). Colour overprints represent successive corrected versions of the curve between 2001 and 2008. The corrections have had the effect of decreasing the earlier temperatures (pre-1970) and increasing later temperatures, which causes an overall increase in the rate of warming between 1850 and 2010. Results are plotted as anomalies, which is the difference between the calculated global average temperature for any one year and the 1961-1990 Climate Normal. The CRU-TEM curve (currently CRU-TEM4) is the global temperature record used by the IPCC.

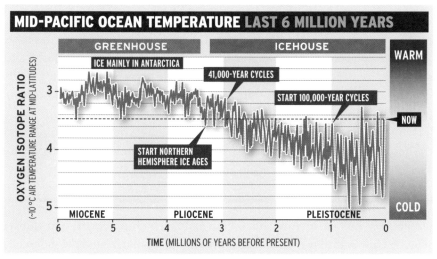

Fig. 2. Pacific Ocean deep water temperature for the last 6 million years, showing greenhouse and icehouse phases of Earth's history (Mix et al., 1995a, b). Dotted horizontal line indicates today's temperature. Based upon oxygen isotope ratios measured from minute fossil shells of seabed-dwelling foraminifera in marine cores. (Foraminifera are single-celled, microorganisms with a tiny chambered shell (up to 1 mm in size) made of calcium carbonate ($CaCO_3$). After death their hard parts accumulate as fossils within ocean floor muds.)

or two only of the many subdisciplines relevant to the climate system and its impacts. As Canadian professors Chris Essex and Ross McKitrick have remarked, 'On the subject of climate change everyone is an amateur on many if not most of the relevant topics.'

It is widely believed that the study of climate change is the exclusive province of meteorologists (who study atmospheric weather systems) and climatologists (who study the longer term averages of weather-related statistics on a monthly to an annual scale). In reality, the very wide range of disciplines and subdisciplines that are relevant to the climate system, and hence to climate change, can be grouped into three main categories. The first group comprises scientists in the fields of meteorology, atmospheric physics, atmospheric chemistry, oceanography and glaciology; these persons mostly study change over the timescale of instrumental measurements only, and are therefore primarily concerned with the atmospheric and oceanic processes that control weather and its variability. A second group comprises geologists and other Earth scientists, who hold the key to delineating climate history and the inference of ancient climate processes. Finally, a third category comprises those persons who study enabling disciplines like mathematics, statistics and computer modelling.

The physical understanding of climate change provided by this already very large group of scientists and disciplines is utilised by a further group of mostly biological scientists, who study the impacts of changing weather and climate on the disposition and evolution of Earth's life forms.

Much of the scientific alarm about dangerous global warming originates with atmospheric scientists (some meteorologists, physicists, chemists) and computer modellers, whose perspective is heavily influenced by their knowledge of daily weather events and extremes. In contrast, climate historians, including many (though not all) geological scientists, see no reason for alarm. This is because of the perspective that such scientists attain from observing the recurring patterns of climate change in the geological record (compare Fig. 2, p.17), which they are made conscious of every time that they inspect an outcrop or a drill core.

Many who study the impacts of weather and climate extremes on both community welfare and on natural flora and fauna claim expertise in climate science. Unfortunately many of these persons fail to understand the difference between the relatively slow changes that occur in the pattern of recurring everyday weather and the impact of rare but severe extreme weather events. Their call to prevent dangerous global warming is invoked under the mistaken belief that such action will prevent the extreme events that are in fact an intrinsic characteristic of both contemporary and past climatic regimes.

Attaining a balanced perspective on climate change requires at least a passing familiarity with all of the three major groups of climate-related specialities, and of the impacts of climate on society and the natural biosphere, a demand that tests even the very best of scientists.

How does the climate system work?
The climate system works through atmospheric and oceanic circulations that continuously transfer excess solar energy from the tropics to polar regions.

The Earth's rotating near-spherical shape, combined with its tilted axis and elliptical orbit, determine that the Sun's energy is not distributed evenly over the whole planetary surface. Notably, only one half of the Earth receives sunlight at any time, with maximum solar radiation intensity occurring when and where the Sun appears directly overhead during the day. Of course, the rotation of the Earth causes this locus of maximum radiation to progressively move westwards (zonally) around the planet during the daily 24 hour cycle.

Maximum mid-day solar radiation intensity occurs in the tropics, with lesser intensity being experienced at progressively higher latitudes towards the poles. Because the axis of rotation is tilted to the plane of the Earth's orbit around the Sun, the latitudinal band of greatest radiation intensity moves north and south (meridionally) between the hemispheres during the seasons of the year, bounded by the tropics of Cancer and Capricorn.

Surprisingly to many people, the changing length of day with season away from the equator results in the daily solar energy received over polar regions in midsummer (when 24-hour daylight pertains) being equivalent to that received in the tropics (where 12-hour daylight pertains). Of course, for similar reasons no solar energy at all is received over the poles in midwinter. When averaged over the annual cycle, it is the heating of the tropics that dominates the climate system, but the summer heating over the poles cannot be ignored.

The strong gradient of annual solar energy input between the equator and the poles requires heat to flow towards high latitudes to maintain overall energy balance (Fig. 3, lower). The energy transports are made by the atmospheric and oceanic circulations (Fig. 3, p.21).

The Sun's energy is largely absorbed at the land and sea surfaces. The absorbed land energy and much of the ocean energy is returned to the atmosphere by direct heat exchange (conduction) and as latent heat from the evaporation of water.[1] The latent heat is made available to the atmosphere as water vapour is condensed into clouds, especially in the deep convection cloud systems of the tropics.

The poleward transport of heat by the atmospheric circulation is largely accomplished by shorter term weather systems, which include the regular passage of atmospheric high and low pressure cells and the frontal systems that accompany them (Fig. 3, p.21). Changes in the rate of transport of heat over long periods are manifest as changing weather patterns and climate.

In essence, therefore, the Earth's climate system is regulated by the absorption of short-wave solar radiation and the subsequent return of balancing longwave radiation to space. Within the atmosphere and ocean, it is surface-atmosphere heat exchange processes, and the distribution of this heat to polar regions by way of convective overturning and larger scale oceanic and atmospheric circulations, that maintain overall energy balance (compare Fig. 15, p.93), and therefore a relatively stable climate regime.

One positive result of these internal processes (from the human perspective) is that the habitability of western Europe is greatly improved by the warmth delivered by the Gulf Stream. Without this and other poleward transfers of heat the higher latitude and polar regions would be much colder in winter than they are today.

What is the Climate Normal?

Climate Normal is a 30-year average of weather, against which longer-term climate change can be assessed.

Weather patterns vary in such a way that it is rare for the meteorological statistics of two successive, comparable seasons to be the same. Generally, the average temperatures will be different, the rainfall totals will be different, and the number and intensity of storm events will be different. Nonetheless, when averaged over many years a (climatic) pattern emerges in which each season's, or the annual, statistics vary about a discernible long term mean.

During the 19th century, as a record of historic weather observations was built up from meteorological stations scattered around the world, scientists

---

[1] Latent energy: heat energy released or absorbed by a substance during a change of state (think evaporation or freezing of water).

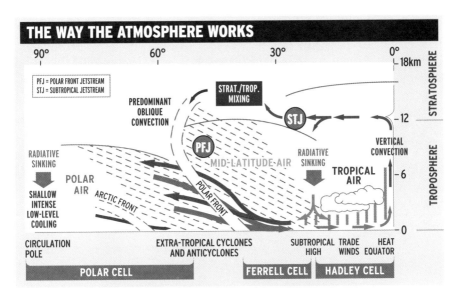

THE WAY THE ATMOSPHERE WORKS

90°   60°   30°   0°
⌐18km

PFJ = POLAR FRONT JETSTREAM
STJ = SUBTROPICAL JETSTREAM

STRAT./TROP. MIXING

PREDOMINANT OBLIQUE CONVECTION

STJ

-12

PFJ

MID-LATITUDE AIR

VERTICAL CONVECTION

RADIATIVE SINKING

RADIATIVE SINKING

TROPICAL AIR

-6

POLAR AIR

ARCTIC FRONT

POLAR FRONT

SHALLOW INTENSE LOW-LEVEL COOLING

-0

CIRCULATION POLE

EXTRA-TROPICAL CYCLONES AND ANTICYCLONES

SUBTROPICAL HIGH   TRADE WINDS   HEAT EQUATOR

POLAR CELL   FERRELL CELL   HADLEY CELL

STRATOSPHERE   TROPOSPHERE

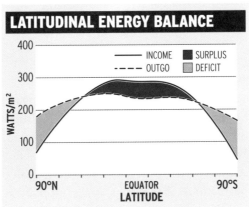

LATITUDINAL ENERGY BALANCE

400

300

200

100

0

WATTS/m²

—— INCOME   ■ SURPLUS
- - - OUTGO   ▨ DEFICIT

90°N   EQUATOR   90°S
LATITUDE

Fig. 3. Upper: Schematic representation of the major features and circulations of the atmosphere (troposphere and lower stratosphere) as a cross-section from the equator (right side) to the pole (left side) (Geerts & Linacre, 1997).

Lower: The effect of tropospheric circulation on the redistribution of solar energy, as summarised on a global transect from the north pole (left) to the south pole (right). Between latitudes 40°N and 40°S the Earth receives more solar radiation than it transmits back to space (red field), whereas at higher latitudes more energy is radiated back to space than is received from the sun (two blue-grey fields). Overall energy balance is maintained by the lateral transfer of excess heat from the tropics towards the poles by the atmospheric (main diagram) and ocean current (not shown) circulation systems.

realised that maximum value could only be achieved from these data if they were systematised. Accordingly, an International Meteorological Organisation (IMO) was formed in 1873, which morphed into the present UN World Meteorological Organisation (WMO) in 1950.

At a meeting in Warsaw in 1935, the members of the IMO agreed on an international standard of comparison by which longer term climate change might be distinguished from shorter term variability, choosing 30 years as the appropriate time over which to average the data.

Applying this criterion, the period 1901–1930 was agreed as being the first Climate Normal. Henceforth, climate atlases presented compilations of Climate Normal averages for different places on the globe, using such meteorological parameters as temperature, humidity, rainfall and pressure.

Thereafter, meteorological statistics could be compared with the local Climate Normal to identify whether a particular month, season or year was above or below the 'Normal', with any departure being regarded as a climate anomaly. The prevailing anomalies from different locations around the world can either be averaged to provide a global picture (Fig. 1, p.17 providing an example), or compared in smaller groups of regions that have similar anomalies, thus allowing their extent and intensity to be judged.

There is nothing unique about a period of 30 years, nor the specific 1901–1930 interval that was chosen for the first Climate Normal. These choices simply reflected the practical recognition that few longer records of systematic observations existed in 1935. Subsequently, the Climate Normal period used for comparison of different temperature records has been updated by the WMO to correspond to, first, averaged 1931–1960 measurements, and more recently to the still-used base period of 1961–1990.

In essence the Climate Normal provides a summary of the averages of climate behaviour at any particular location that defines an envelope of normal variability against which extreme events can be judged. Such base-line information is vital knowledge for persons and organisations charged with planning and management of climate-sensitive societal activities. These include especially land-use and water availability, but also extend to essential infrastructure such as housing, transport and communications.

The resilience of modern, industrialised societies is largely based upon their ability to withstand the impacts of weather similar to that of the Normal, because that is what our infrastructure has been designed to cope with. However, we have much less ability to withstand without damage the spasmodic extreme weather and climate events that occur, and which lie far from the Normal.

Knowledge of the Climate Normal, and the typical variations that occur about it, is therefore a vital aid towards managing climate hazard.

Is there such a thing as a global average temperature?
Yes, but it is difficult to assess and its usefulness is not established.

The powerful term 'average' is much used in discussion about climate change. Public discussion often concentrates on perceived deleterious impacts associated with changes in properties such as 'global average temperature' or 'global average sea-level'.

Such averages are difficult to assess and have no physical existence, but represent instead convenient statistics that are generated from many separate pieces of data gathered from disparate places. In the case of temperature, it is not the measured values that are compared but instead globally averaged anomalies of temperature (see above: What is the Climate Normal?). Thus the commonly referred to nominal 0.8°C surface temperature increase during the 20th century actually represents the change in global averaged annual anomalies over the century.

It is for this reason that there is so much disagreement about the accuracy of various different estimates of temperature and sea-level history. For the construction of these averages requires data to be selected, corrected and statistically manipulated, activities which may quite legitimately be undertaken in different ways by different investigators and which then may lead to slightly different outcomes.

For example, the Hadley Centre of the UK Meteorological Office publishes a record of the monthly and

annual global temperature anomaly since 1850 (Fig. 1), and the Goddard Institute of Space Studies (GISS) of NASA publishes another. Though similar overall, these two records differ in fine detail. That both organisations amend their earlier assessments from time to time indicates an acceptance that there is an inevitable uncertainty associated with such assessments, no matter how carefully they are performed.

Accepting that, nonetheless a global average temperature statistic can obviously be calculated using results from many individual weather stations. Some scientists argue that such a number can have no more meaning than does a global average telephone number. The fact you can calculate such numbers does not, per se, confer any deep meaning or usefulness upon them. Importantly, while changes in regional temperature contribute towards changes in global average temperature, a change in global average temperature tells us nothing about the regional patterns and differences in temperature that have led to the changed average. This is important when we consider the potential impact of changing global average temperature on regional climate.

Although temperature is an important property for most activities that are related to living organisms, it is not an intrinsic property of materials, i.e. you cannot add or subtract temperature to a substance. Rather, temperature changes occur because of changes in the amount of heat (which is an intrinsic property) present. In essence, the temperature of a substance represents the outcome of the various energy exchange processes that are going on at any moment.

In any case, real-world environmental effects are not imposed by changes in global average conditions but by changes in specific local conditions. What is of concern to the citizens of different cities and farmers around the globe is whether their own local temperature, rainfall or sea-level are going up or down, not what conceptual global averages might be doing.

How do climate and weather differ?
Climate is a long-term outcome of transporting solar energy around the Earth; weather comprises the atmospheric processes that help to achieve that transport.

For good statistical reasons, meteorologists have chosen to define climate as representing a 30-year average of weather — expressed, for example, in terms of an average of the annual temperature or rainfall over 30 years at a particular location (see above: What is the Climate Normal?). The ensemble of monthly, seasonal and annual weather patterns will not depart too far from the Climate Normal.

The underlying concept was well expressed by Mark Twain, who coined the aphorism, 'Climate is what you expect; weather is what you get.'

People intuitively understand this saying. All of us know that the vagaries of the weather provide a daily smorgasbord of change that controls our outdoor activities, the discussion of which serves as a ritualised tool for social small talk even amongst strangers.

On the longer time scale people also understand that the seasons come and go, that groups of years that are hotter or cooler, or wetter or drier, irregularly present themselves from time to time, and that such natural variability (which regularly imposes environmental and social consequences) does not necessarily indicate a change in climate.

However, and for obvious reasons, people mostly have little appreciation for environmental changes that occur over periods longer than a human lifespan. Such longer term (climatic) changes generally occur on a time scale of decades to centuries to millennia to millions of years, and have been identified in many geological studies from around the globe.

To add to the complexity of our understanding, studies of ancient environments show that marked climatic change (such as the change of several degrees in global average temperature that accompanies a glacial or an interglacial episode) can also occur abruptly over periods as short as a few decades.

Mother Nature, then, does not recognise the arbitrary distinction between weather and climate that human psychology finds so useful. Instead, basic physical and chemical mechanisms operate over all time scales and result in a continuum of change. It is human perception and convenience only that distinguishes between weather and climate, for both are attributes of the overall climate system.

The one constant about weather and climate is that change occurs constantly and on all time scales, and does so in a chaotic fashion that is generally unpredictable beyond the limits of several-day weather forecasting. As the IPCC said, memorably and accurately, in their 3rd Assessment Report in 2001: 'The climate system is a coupled non-linear chaotic system, and therefore the long-term prediction of future exact climate states is not possible'.

Is there such a thing as a global climate?
Notionally, yes; but mainly for technical scientific use at a high level of abstraction.

Early in the 20th century, the Russian geographer Wladimir Koeppen demonstrated that the world could sensibly be subdivided into 28 climatological-vegetational zones that ranged from polar, through temperate and arid, to equatorial conditions. The demarcation into forest, grassland and desert, combined with temperature and rainfall, corresponds to regimes that are natural. For example, vegetation does not grow profusely when there is low rainfall, irrespective of whether the temperature is polar, temperate or tropical. Koeppen's orginal zones were so well chosen that they are still recognised by modern geographers.

Local climates differ from place to place on Earth because of changing geography and topography. Nonetheless, characteristic latitudinal bands of similar climate occur, and these reflect the role of overturning atmospheric circulations in the transfer of heat from the tropics (compare Fig. 3, p.21). Also there are recurring climate types in both hemispheres where similar land-sea arrangements occur, for example the 'mediterranean climates' typical of west coast locations in temperate latitudes.

It is, of course, possible to take an average of all Koeppen's zones and declare that to be the global average climate. But such a level of abstraction has only limited use, because it deliberately reduces the amount of meaningful information under consideration. Humans do not live in a global climate, but rather in their own specific local or regional climate; and the same is true for the natural organisms around us, each species of which is specifically adapted to a particular set of environmental conditions.

Thus global climate exists only as a notional average of all local and regional climatic states, which is a concept of limited value. Our concerns about climate change and climate hazard, therefore, should not centre on changes in global climate averages but on changes in the specific locations where large numbers of people live.

Is the climate changing?

Climate is always changing: despite wide currency, the phrase 'climate change' is a tautology.

Despite the limitations of concepts such as global average temperature, modern climate change is generally described by referring to statistically-averaged records based upon temperature measurements collected at a worldwide network of meteorological stations.

For temperature, this immediately restricts the discussion to changes that have occurred over only the last 150 or so years (Fig. 1, p.17). Thus the discussion becomes largely about a changeable and changing weather history that in total represents a span of five Climate Normals, i.e. just five climate data points (see above: What is the Climate Normal and How do climate and weather differ?). Given year to year variability, the question that arises is whether the small increase in average temperature that has occurred through the 20th century has had an important effect on daily life.

To assess properly whether climate is changing in the long term requires the study of geological records of the changing physical and chemical composition of materials that accumulated over ancient times past. Suitable physical samples for study occur in layered sediment deposits from the floor of lakes and the oceans, and snow (ice) accumulations on land; typical biological climate indicators include tree rings and coral growth rings (see III: What is a proxy record of temperature?). Importantly, such records allow for the primary reconstruction of only relative climate variations, the actual magnitude of which must be determined using modern comparison or experimental correlations. There is thus always a degree of uncertainty as to the precise timing and magnitude of past climatic variations.

Analysis of the last 6 million years of climate history from ocean seabed sediment cores reveals that prior to 3 million years ago temperature was generally 2-3° C warmer than today (Fig. 2, p.17), a period that has been called the Pliocene Climatic Optimum. Since then, a global cooling trend has been in progress.

The cooling trend was initially accompanied by an increasing magnitude of the small 41,000 year temperature oscillations that occur throughout the geological record. As the amplitude of this oscillation increased, the modern

100,000 year glacial and interglacial cycle emerged too, and came to dominate especially during the last 0.5 million years. For most of the last million
years Earth has experienced glacial conditions, interrupted by only brief interglacial periods of warmth such as the one that we enjoy now. Deep ice
cores drilled through the ice sheets of Antarctica and Greenland confirm the
existence of, and have provided much detail about, the most recent 100,000
year-long glacial-interglacial cycle.

Just 20,000 years ago, the Earth was experiencing the last glacial maximum
(Fig. 4, p.28). Great ice sheets covered North America to between the 40th
and 50th parallels of latitude; a similar ice sheet covered North West Europe
and the region of the Alps. The source of the ice was a transfer of water by
evaporation from the oceans and precipitation of snow at high latitudes, as a
result of which global sea-level was about 130 metres lower than now. Temperatures during the last glaciation were up to 20°C cooler than today over
Greenland and 10°C cooler over Antarctica. In contrast, the tropical ocean
surface temperatures of the western equatorial Pacific Ocean are estimated to
have been only about 2–3°C cooler then. The warming from the last glacial

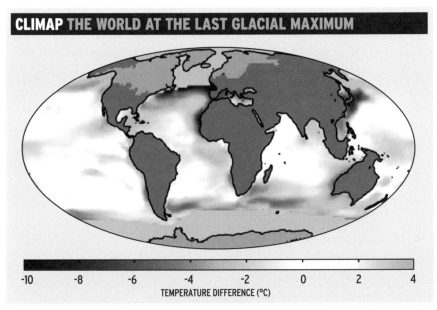

Fig. 4. Reconstruction of transformed global geography at the time of the last glacial maximum,
20,000 years ago (CLIMAP research program). Note the large ice-cap up to 3 km thick in
the northern hemisphere (grey); the greatly expanded sea-ice in the Arctic Ocean and around
Antarctica (lighter grey); the increased landmass areas around the edge of the continents caused
by a global sea-level drop of about 125 m (brown, including a greater Australia); and the changes
in average temperature of the ocean, generally cooling by several degrees (blue shades) but
possibly with some areas of mid-latitude warming (yellow).

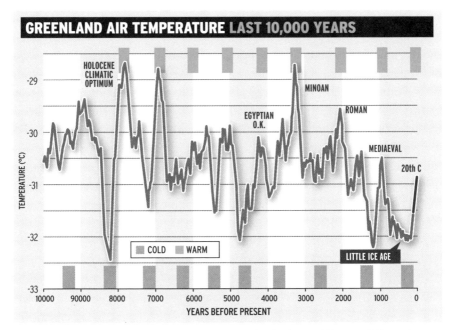

Fig. 5. Greenland surface air proxy-temperatures for the last 10,000 years (Holocene) as reflected in palaeo-temperatures derived from changes in oxygen isotope ratio in the GISP2 ice core (Alley, 2000). Short warm periods, like the Medieval Warm Period and Late 20th Century Warming, occur about every 1,000 years (pink bars, top), separated by cool periods such as the Little Ice Age (blue bars, bottom). This pattern, called the Bond Cycle and probably of solar origin, is superimposed on a cooling trend of about 0.25°C/thousand years since the Holocene Climatic Optimum (HCO). Note that the Minoan, Roman and Medieval Warm Periods, and the HCO, were all significantly warmer than the 20th century warming.

period began about 18,000 years ago and was completed by 10,000 years ago (compare Fig. 6, p.31).

Overall, the pattern of climate variation over the past 6 million years has been one of cooling and recovery, with cooling episodes becoming successively colder but each time recovering to near earlier temperatures. However, the current Holocene interglacial is not quite as warm as were previous, relatively recent interglacials. For example, the last interglacial, about 125,000 years ago was up to 2°C warmer than today (Fig. 2, p.17), causing a reduced-size Greenland icecap, little sea ice in the Arctic Ocean and a global sea-level several metres higher than at present.

The current interglacial is called the Holocene.[2] While not as warm as the last interglacial, the Holocene has now lasted a little longer than the 10,000

---

[2] The Holocene is the geological name given to the period of time that has elapsed since the end of the last glaciation about 11,700 years ago (Fig. 6, p.31). The Holocene therefore represents a warm (interglacial) period. The early and middle parts of the Holocene correspond to the human Mesolithic and Neolithic cultural periods. See also III: What is the Holocene and why is it important?

years that represent the average length of recent interglacials. Despite this fact, and despite the long-term gentle cooling that has occurred since the Holocene Climatic Optimum (Fig. 5, p.29), no unambiguous indicators yet exist of accelerated cooling towards the next glaciation.

It is clear, then, that the geological record provides abundant evidence that climate always changes. In this context claims that Earth's climate has been stable for the last two millennia, until human carbon dioxide emissions caused the mild warming that occurred during the 20th century, are simply unsustainable against strong evidence to the contrary. Alarmist assertions about dangerous climatic warming being in progress are based on statistics that both suppress the magnitude of earlier climate variability and also enhance the late 20th century warming. Dangerous AGW proponents also fail to heed the large corpus of scientific, cultural and anthropological evidence that conflicts with their belief that dangerous warming is already occurring.

Any sensible narrative on climate policy has to start with the observation that climate has always varied over a range of timescales, often for reasons that we do not fully understand. The minor 20th century warming took place against a long-term cooling trend since the Holocene temperature optimum, and today's temperatures are therefore in no way unusual. Other similar warming episodes of a degree or so, with about a 1,500 year separation, occurred during the Egyptian, Minoan, Roman and Medieval warm periods — each period of warmth since the Minoan one being slightly cooler than its predecessor (Fig. 5, p.29).

The late 20th century warming, and any renewed warming that may occur this century, should therefore be welcomed rather than feared. For the alternative of a colder and drier climate, and the extension of polar ice sheets and mountain glaciers as Earth passes into the next ice age, entertains a far worse prospect for humanity.

Is today's temperature warming or cooling?
How long is a piece of climate string? Whether we perceive a warming or a cooling trend depends upon the period of time chosen.

Endless public discussions have been conducted over the last 20 years about whether the global temperature is rising or falling, based solely upon analysis of historic thermometer measurements of temperature (compare Fig. 1).

These discussions lack proper scientific context and cast many among our commentariat as akin to medieval priests obsessing about angels and pins. For, of course, the question in hand cannot be answered by poring minutely over a record of temperature a mere 150 years long.

Whether today's temperature is unusually warm can only be judged against knowledge of previous variations in temperature and their causes. Unfortunately, climatic records prior to the last 150 years, which are based upon proxy measurements, do not provide direct knowledge of ancient global temperature, but reflect instead local or regional climate histories, to which we now turn.

One of our highest quality long climate records comes from the Greenland GISP2 ice-core, measurements from which have allowed reconstruction of the temperature back into the last ice age (Fig. 6). The results show that the peak of the last glacial cold period occurred about 20,000 years ago, after which rapid warming occurred up until the start of our modern interglacial warm period, 11,700 years ago. Thereafter, in the early Holocene about 8,000 years ago, temperatures were more than 2°C warmer than today.

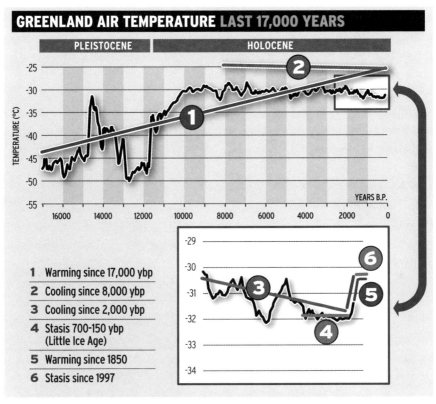

Fig. 6. Greenland surface air proxy-temperatures for the last 17,000 years as reflected by changes in the GISP2 ice core, based upon oxygen isotope ratios (Alley, 2000). Numbered, colour trend lines indicate the direction (warming or cooling) and rate of temperature change (line slope) over various time periods (Davis & Bohling, 2001). Note that temperature has warmed since 17,000 and 150 years ago (red), cooled since 10,000 and 2,000 years ago (blue) and remained static (neither warming nor cooling, on average) (green) since 700 years ago and also (from Fig. 9, p.75) since 1997.

Our initial question about whether warming or cooling is occurring today can now be addressed by inspecting the graph of reconstructed temperature for Greenland (Fig. 6, p.31). It is apparent that warming has taken place since just after the last glaciation, from 17,000 years ago, and also since 160 years ago. But over intermediate time periods both cooling and temperature stasis have occurred, with cooling trends since 8,000 and 2,000 years ago, and stasis between 700–150 AD and 1997–2012 AD.

The instrument record of warming over the last 150 years is a continuation of warming present near the top of the Greenland ice-core record. The GISP2 temperature record, and other similar records from around the world, shows that the 20th century instrumental warming represents a second-order climatic fluctuation on the long-term cooling trend established since the Holocene Climatic Optimum.

The clear conclusion is that the answer to the question, 'is global warming occurring' depends fundamentally on the length of the piece of climate string that you wish to consider.

Is today's temperature unusually warm?
No — and no ifs or buts.

It follows directly from the questions already discussed, and fashionable media opinion notwithstanding, that modern temperatures are not unusually warm.

In recent geological history alone, temperatures were up to 2-3°C higher than today for several million years during the Pliocene period (6-3 million years ago; Fig. 2, p.17), for briefer periods during recent warm interglacials (including the early Holocene) and during the historic Minoan, Roman and Medieval Warm Periods (Fig. 5, p.29).[3]

For emotional impact, it is often stated that current temperatures are the warmest for nearly a thousand years, a statement that may be more or less true given the length of the Little Ice Age. However, the choice of a one thou-

---

[3] The Medieval Warm Period (MWP) corresponds to a worldwide benignly warm period, 850-1100 AD, during which, for example, grapes were able to be grown in the north of England.

sand year period to judge climate change against is entirely arbitrary, and overlooks the warm period that occurred prior to the 14th century inception of the Little Ice Age[4] (III: What were the Medieval Warm Period and Little Ice Age?) and also similar earlier warm periods.

Viewed in the proper time context, therefore, the warmer temperatures at the end of the 20th century simply indicate the occurrence then of the proximity of a cyclic millennial peak within Earth's normally and naturally varying climatic history.

How much warming actually occurred in the 20th century?
Somewhere between 0.4°C and 0.8°C.

Until recently, it has been generally accepted that a planetary warming of 0.7°–0.8°C occurred during the 20th century. Dangerous warming proponents assert that this warming was mostly of anthropogenic cause; that is, associated with emissions of carbon dioxide caused by the industrial activity of modern societies.

However, independent scientists have long suspected that the apparent warming figure of 0.8°C was inflated by the urban heat island effect, whereby thermometers situated within and in the vicinity of large towns and cities (the majority of monitoring sites) have their readings inflated due to the changing nature of their surroundings. In particular, as vegetation is replaced by buildings, paving and roads the natural cooling effect of evapotranspiration is lost and local temperature rises. In addition, independent analyses have suggested that adjustments made to individual temperature records within national meteorological agencies have tended to exacerbate the recent apparent warming trend.

In a paper presented at the European Geosciences Union in 2012, Greek scientists Steirou and Houtsoyiannis showed that this is indeed the case. Their estimate is that for 67% of the 181 globally distributed weather stations that they examined, adjustments had been made that resulted in 'increased positive

---

[4] The Little Ice Age (LIA) roughly spanned the years 1350-1860 AD, during which time several lows in global temperature were about a degree cooler than they are today. Several of these lows coincided with the declines in solar intensity known as the Maunder (1645-1715) and Dalton (1795-1840) sunspot minima, which were accompanied by widespread cold, famine and starvation in Europe.

[temperature] trends, decreased negative trends, or changed negative trends to positive', whereas by chance alone the expected proportions of stations recording extra adjusted warming should have been one out of two, or 50%.

Steirou and Houtsoyiannis conclude that their results, 'tend to indicate that the global temperature increase during the last century is between 0.4°C and 0.7°C, where these two values are the estimates derived from raw and adjusted data respectively'.

But the UN says that '13 of the warmest years ever have occurred in the last 15 years'.
The truth of this claim depends upon what 'ever' means: but even were it to be true, so what?

Variations on this statement are rampant in the media and give weight to the assertion that dangerous global warming is being caused by human greenhouse emissions. Yet persons who repeat this claim thereby highlight only their innocence of knowledge of the science of climate change.

The key word, the misappropriation of which confers an element of truth on the statement, is 'ever'. To dangerous AGW proponents, this word means 'since accurate instrumental temperature records began', i.e. over about the last 150 years (e.g. Fig. 1).

But as we have already discussed several times, 150 years is a trivially short and inadequate period over which to make judgements about climate change.

Furthermore, as the last 150 years has fortuitously seen the Earth pass out of a Little Ice Age, it is scarcely surprising that temperatures have become slightly warmer during the 20th century. In addition, the Little Ice Age and the 20th century warming correspond, respectively, to the colder and warmer parts of the natural 1,000 year-long climate rhythm (Fig. 5, p.29).

'Ever' does not mean 'the last 150 years', and for the IPCC to equate 'ever' with the recent instrument record of climate is misleading, whether or not that was the intent. For, to repeat an earlier conclusion, viewed in proper geological context there was nothing unusual at all about the warmth of planet Earth at the end of the 20th century.

Is dangerous global warming being caused by human-related carbon dioxide emissions?

No evidence exists for this proposition.

The scientific rationale for the proposition that 20th century warming is caused by human carbon dioxide emissions is that carbon dioxide is a greenhouse gas and any enhancement of its atmospheric concentration will cause further and dangerous warming. Though there is a kernel of truth in the first part of the proposition, it has never been demonstrated that warming above today's temperature would be harmful.

For example, biological diversity has expanded and life on Earth has flourished over several hundred million years, including times when temperatures were warmer and conditions more humid than now. Modern equatorial rainforests support a rich diversity of life, and in general biodiversity is enhanced rather than diminished in moist, hot climes.

However, the real issue is not so much that additional carbon dioxide will enhance the greenhouse effect per se, but rather how much such enhancement will occur. Scientists usually discuss this problem within the context of the sensitivity of Earth's temperature to a doubling of atmospheric carbon dioxide

The IPCC has given support to the hypothesis that accumulating human-related emissions of carbon dioxide will lead to dangerous warming, by drawing attention to the correlation between a rising concentration of carbon dioxide and rising global temperature during the 20th century. It is claimed that most of the presumed 0.4°C warming of the second half of the 20th century, (Fig. 7, p.36) when atmospheric concentration of carbon dioxide rose from 300 to 350 ppm, resulted from human-related emissions.

This logic is valid in principle, for carbon dioxide is a greenhouse gas and increasing its concentration will cause warming (or, in particular circumstances, turn a natural cooling trend into a less strong cooling trend). However, the crux of the discussion is the sensitivity of global temperature to increasing carbon dioxide (IV: What is climate sensitivity?). That is, whether the increase in carbon dioxide from 300 to 390 ppm that has occurred since 1950 is sufficient to have caused the 0.4-0.7° of warming measured over that time.

Enhanced public concern has been raised about this issue, without rigorous science to back the claim, by some scientists asserting that further warming will exceed an hypothesised 'tipping point' leading to runaway (or at the least irreversible) global warming.

The first point to make is that, even were they to have been caused entirely by human-related emissions, the rates of warming of up to about 1.7°C/century that occurred during the two 20th century warming pulses century

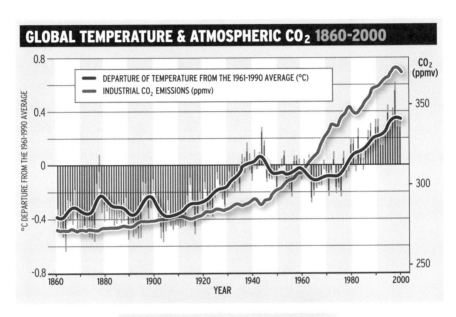

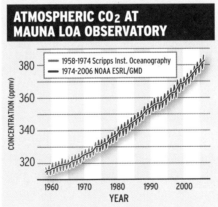

Fig. 7. Atmospheric surface temperature anomaly curve for 1860-2000, as reproduced in the IPCC's 3rd Assessment Report . Superimposed blue line represents the graph of human-related carbon dioxide emissions. Note that the global temperature cooled for 30 years at the very time (1940-1970) that industrial carbon dioxide emissions were rising most rapidly. Inset: carbon dioxide levels as measured at Mauna Loa, Hawaii for 1952-2000. Note the presence of regular seasonal variations.

(1910-40 and 1975-98) are neither unusual nor dangerous. Rather, they fall well within the typical rates of natural warming and cooling of up to $\pm$ 2.5°C/century that have occurred throughout the last 10,000 years.

Second, climate warming phases that occurred during the Holocene or earlier were not always, or even often, accompanied by increases in atmospheric carbon dioxide. Conversely, the large increase in carbon dioxide caused by post-World War II industrial expansion between 1940 and 1976 was accompanied by temperature cooling, not warming (Fig. 7, p.36), and neither has warming occurred since 1997 despite a parallel 8% increase in carbon dioxide.

Third, the magnitude of the most recent half-century warming of perhaps 0.4°C is less than the magnitude of year-to-year natural variability of the global temperature record. For example, the warming and cooling associated with the El Niño-Southern Oscillation phenomenon (an internal variability of the climate system; see VII: What is ENSO and how does it affect Australian climate?) regularly produces temperature changes of larger magnitude than this.

Fourth, because the extra warming caused by adding carbon dioxide to the atmosphere diminishes in magnitude with increasing gas concentration, less warming is caused by each successive increment in carbon dioxide (IV: Is less warming bang really generated by every extra carbon dioxide buck?) Therefore, most of the impact of carbon dioxide on Earth's global temperature was produced by the first 50 ppm of concentration of the gas in the Earth's atmosphere, and further increases in concentration beyond that have exercised, and will continue to exercise in future, a diminishing incremental increase in global temperature. The concept of a 'tipping point', beyond which global warming accelerates, is not supported by any known science.

If earlier warming events in geological history cannot be attributed to increased carbon dioxide, then it is not logical to attribute the recent warming to a coincident increase in carbon dioxide. The frailty of logic is further exposed when it is realised that, during the 20th century, it is only the period 1975–98 that rising temperature and strongly rising carbon dioxide were coincidental. Relatively little increase in carbon dioxide occurred during the 1910–1940 warming period; and temperatures rose neither during the 1940–1975 post-World War II industrialisation period of high carbon dioxide emissions, nor over the last 16 years when a further 8% increase occurred in carbon dioxide.

It is therefore more likely that the recent warming is part of the patterns of natural change than that is has been driven by increased atmospheric carbon dioxide. Natural agents of temperature change include climatic oscillations such as ENSO and the Pacific Decadal Oscillation (compare Figs. 30, 31, pp.156-7), and also the longer cycle associated with Earth's rebound from the Little Ice Age (compare Fig. 42, p.232) — the effects of none of which are included in the computer models that are cited in support of dangerous AGW.

To sum up, although atmospheric carbon dioxide concentration has increased since industrialisation, and especially over the last half century, it is simply not possible to isolate any human enhancement of the greenhouse effect from temperature changes that may have been driven by natural climate variability.

II

The Meat and Eggs
of Climate Alarmism

What set off climate alarmism?

Events way back when: but modern global warming alarmism gained momentum in the 1980s.

One of the earlier recorded comments about climatic warming was made in 1817, in a report to the British Admiralty by the president of the Royal Society of London:

> *It will without doubt have come to your Lordship's knowledge that a considerable change of climate inexplicable at present to us must have taken place in the Circumpolar Regions, by which the severity of the cold that has for centuries past inclosed the seas in the high northern latitudes in an impenetrable barrier of ice has been during the last two years greatly abated. Mr. Scoresby, a very intelligent young man who commands a whaling vessel from Whitby observed last year that 2000 square leagues[5] of ice with which the Greenland*

---

[5] 22 million square kilometres [1 league = 3.5 miles = 5.6 km].

Seas between the latitudes of 74° and 80°N have been hitherto covered, has in the last two years entirely disappeared.

Ironically from the present-day perspective, though sensibly from the Admiralty's point of view then, the president actually welcomed the warming because sea ice would be melted and navigation passages opened up.

More generally, media mentions of climate change, be they about warming or cooling, are almost invariably couched in alarmist terms. For example, on June 24, 1874, *Border Watch* noted that a curious change in the climate of Scotland was having a deleterious effect on yields of fruit and vegetables, commenting:

All is changed or changing now, although several winters of late years have been remarkable for their mildness, and proved most favourable for flowering plants. The Scotch, however, cannot feed on flowers, and are much to be pitied under the calamity with which they are threatened, of being dependent on our English greengrocers and fruiterers for their supplies of fruit.

Truer words have surely seldom been written, for it is indeed hard to imagine your average Scotsman being happily sustained on flowers.

At the turn of the century and just after, concern passed on to the circumstances of heat waves and droughts that were occurring in both Australia and America. Thus the *Clarence and Richmond Examiner* reported on August 24, 1911, that:

An unprecedented wave of heat rolled over the greater part of the United States during the first five days of July, causing hundreds of deaths, and loss to the growing crops running up into the millions. The mortality in the larger cities was greater than that of the entire Mexican revolutionary war. At least 750 deaths are to be directly attributable to the heat wave, according to the reports telegraphed from the cities affected.

The disastrous wave of heat subsided on July 6, cooling rains falling throughout a considerable portion of the country. This saved the corn crop from utter annihilation. As it was, very serious damage was caused on the farms, products of all kinds suffering. The Illinois State Board of Agriculture, in a special report, says the heat spell has badly damaged all the crops in that State, some of them irreparably. Official reports show that the fruit crop of Iowa fell off 10 per cent in five days.

A Chicago theorist has advanced the idea that the heat generated by the great cities at the present day is changing their climates to a marked degree. Observations covering many years, he says, demonstrate that the climate of New

York has become both warmer and drier with the growth of the city. The rainfall has dropped in recent years from an average of 45in. to 40in.

The ensuing first half of the 20th century, up to as late as the 1950s, saw widespread media concern about untoward warming, especially in high latitudes of the northern hemisphere. In a typical example, the Adelaide Advertiser wrote in April, 1923:

> Reports from fishermen, seal hunters, and explorers who sail the seas around Spitzbergen and the eastern Arctic all point to a radical change in climatic conditions, with hitherto unheard-of high temperatures on that part of the Earth's surface … Many of the old landmarks are greatly altered, or no longer exist. Where formerly there were great masses of ice, these have melted away, leaving behind them accumulations of Earth and stones such as geologists call 'moraines'. At many points where glaciers extended far into the sea half a dozen years ago they have now entirely disappeared … This state of affairs is a cause of much surprise and even astonishment to scientists, who wonder whether the change is merely temporary or the beginning of a great alteration of climatic conditions in the Arctic, with consequent melting of the polar ice sheet.

What we might view as modern climate doomsday writing started in the 1970s, provoked by a decrease in temperature that some scientists feared might mark the start of the next ice age. Colourful English astronomer Fred Hoyle, among others, even wrote a book about the threat entitled *Ice: The Ultimate Human Catastrophe*, and many newspapers reported the issue. The view of *Newsweek* on April 28, 1975, was:

> There are ominous signs that the Earth's weather patterns have begun to change dramatically and that these changes may portend a drastic decline in food production — with serious political implications for just about every nation on Earth … A survey completed last year by Dr. Murray Mitchell of the National Oceanic and Atmospheric Administration reveals a drop of half a degree in average ground temperatures in the Northern Hemisphere between 1945 and 1968 … And a study released last month by two [other] NOAA scientists notes that the amount of sunshine reaching the ground in the continental U.S. diminished by 1.3 per cent between 1964 and 1972.

During the 1970s, as it is sometimes wont to do, the global temperature trend turned around, and warming proceeded from 1979 through to 1998. Starting in the early 1970s, scientists began using computer models to predict the behaviour of the climate system, which led environmentalists to question whether human-related carbon dioxide emissions might not cause dangerous global warming. This idea rapidly gained currency, leading, first, to the convening of an important meeting at Villach, Austria in 1985; and, second, to the creation of the Intergovernmental Panel on Climate Change (IPCC) under the umbrella of the United Nations in 1988.[6]

The various quotations above reveal that where climate change is concerned, there is nothing new under the human sun. But from the viewpoint of modern climate politics, the formation of the IPCC in 1988 enshrined global warming alarmism at the highest intergovernmental and international levels, where it remains in play to this day.

What role did UNEP and WMO play in the creation of the IPCC?

They were the co-founding organisations of the IPCC in 1988.

The climate modelling experiments of the early 1970s cemented the issue of man-made global warming, which had first been raised by Swedish scientist Svante Arrhenius in 1898, as a potential environmental hazard. Consequently, dangerous AGW became the common theme of a series of international and intergovernmental conferences on environmental protection held at that time. These discussions, and growing concerns about the effects of modern industrialisation on the environment, led to the formation of the United Nations Environment Programme (UNEP) at a Conference on the Human Environment in Stockholm in 1972.

A little later, in 1979, the First World Climate Conference held in Geneva alerted the world community to the need for a better understanding of climate systems and climate change. Global cooling, and the possibility of Earth slipping into the next ice age, was the dominant theme at the time. However, the issue of dangerous AGW also received attention. This led to the convening in 1985 of a critically important conference at Villach, Austria, which reviewed the impact of human-related carbon dioxide emissions on climate.

---

[6] Throughout the book, statements are made about the amount of global warming that has occurred over several different historical time periods, depending on the context of the discussion at a particular point. This can be confusing, so readers may wish to note that the following statements are all true (cf., Figs. 1, 7, 10): (i) about 0.9° C of warming has occurred since the end of the LIA in 1860; (ii) about 0.7-0.8° C of warming occurred during the 20th century, in two main periods of ~0.4° C in 1910-1940 and 1975-1998; and (iii) no significant warming has occurred since 1997, i.e. for the last 16 years.

The Villach conference was mainly instigated by UNEP, but importantly for its scientific credibility two co-sponsors were the World Meteorological Organisation (WMO) and the International Council of Scientific Unions (ICSU). After the presentation of invited papers, a forthright Conference Statement was fashioned which included the claim:

As a result of the increasing concentrations of greenhouse gases, it is now believed that in the first half of the next century a rise of global mean temperature could occur which is greater than any in man's history.

The Villach Statement became the launching pad for strong national and international efforts to raise community awareness to the potential dangers of burning fossil fuels and rising atmospheric carbon dioxide concentrations. Lead players in Australia at the time were the Commission for the Future and CSIRO, who sponsored a conference of invited scientists in December, 1987 titled Greenhouse: Planning for Climate Change.

An underlying assertion from the Villach conference, which is still promulgated to this day, was that planning for the future cannot be based on historical data because human-caused emissions of carbon dioxide are now contributing to global warming and climate change in addition to natural causes. And so were politicians delivered into the hands of the computer modellers from whose clutches they yet have to escape, and whose models were then projecting a rise of 1.5-4.5°C in temperature and a 20-140 cm rise in sea-level for the anticipated doubling of carbon dioxide in the atmosphere.

The very strong political promotion of man-made global warming by UNEP and the environmental movement that followed the Villach conference became of concern to the more conservative, science-orientated WMO. First, and as they still do today, the policy proposals were running far ahead of perceived scientific understanding; and, second, the lead in climate matters, a scientific issue, was being usurped by a political organisation, UNEP.

These tensions were resolved in the short term by WMO and UNEP agreeing that a thorough and continuing review should be carried out of the science associated with possible human-caused warming. In 1988, therefore, the two agencies co-sponsored the formation of the Intergovernmental Panel on Climate Change (IPCC) under UN auspices, with the intention that the new organisation should become the authoritative source of advice to governments on climate change issues.

What is the IPCC?

An arm of the UN charged with reviewing the influence of atmospheric carbon dioxide on climate.

Contrary to public perception, the IPCC does not examine the full array of influences that affect climate and climate change. Instead, the IPCC's Charter directs the organisation to assess peer-reviewed research that is 'relevant to the understanding of the risk of human-induced climate change'. Thus the primary functions of the IPCC are to assess the role of human-related carbon dioxide emissions in modifying global climate, the likely impact this

might have on human society, and what responses society might take to mitigate those impacts.

The IPCC operates alongside another UN instrument, the 1994 Framework Convention on Climate Change (FCCC). In Article 1.2, this Convention states that:

> *Climate change means a change of climate which is attributed directly or indirectly to human activity that alters the composition of the global atmosphere and which is in addition to natural climate variability observed over comparable time periods.*

The restricted objectives of both the IPCC and the UNFCCC have caused the UN and its scientific advisers to show little interest in the reasons for, and the possible extent of, natural climate variability. Therefore, and as not widely appreciated, the IPCC only advises governments on the narrow task of assessing climate change that is related to human greenhouse gas emissions. It is easy for bias to develop in such circumstances, wherein all changes that cannot be unequivocally assigned to another cause become attributed to carbon dioxide.

The declared intention of the IPCC was to provide disinterested summaries of the state of climate science as judged from the published, refereed scientific literature. In reality, the four successive Assessment Reports in 1990, 1996, 2001 and 2007 (all available at the IPCC web site: *http://www.ipcc.ch/*) have promulgated an increasingly alarmist view of human-caused warming (Table I). At the very same time, the evidence for a human influence has been weakening, and the more balanced views on the issue of many qualified independent scientists have been marginalised or ignored.

Today, advice from the IPCC is the linear thread that underlies all national and international efforts to control the emission of greenhouse gases. But as outlined in many recent writings, including a compulsively readable book by Donna Laframboise (*The Delinquent Teenager Who Was Mistaken for the World's Top Climate Expert*), the IPCC's unbalanced brief and partial reporting has inevitably led to advocacy research and the provision of dangerously inadequate information to governments about climate change.

IPCC masthead advice, 1990-2012

The observed [20th century temperature] *increase could be largely due to … natural variability. (IPCC, 1st Assessment Report, 1990)*

The balance of the evidence suggests a discernible human influence on climate. (IPCC, 2nd Assessment Report, 1995)

There is new and stronger evidence that most of the warming over the last 50 years is attributable to human activities. (IPCC, 3rd Assessment Report, 2001)

Most of the observed increase in globally averaged temperature since the mid-20th century is very likely [=90% probable] *due to the observed increase in anthropogenic greenhouse gas concentrations. (IPCC, 4th Assessment Report, 2007)*

It is extremely likely [>95% probable] *that human activities have caused more than half of the observed increase in global average surface temperature since the 1950s. (IPCC, 5th Assessment Report, DRAFT, 2012)*

Table 1. Masthead summary statements about human greenhouse warming published by the IPCC in Assessment Reports 1-4, and in the draft text for Report 5 (to be released in late 2013)

What is the NIPCC?
An independent scientific endeavour called the Non-Governmental International Panel on Climate Change.

The NIPCC was initiated in 2008 by Professor Fred Singer, a former Director of the U.S. Satellite Weather Service and a doyen of meteorological researchers. The NIPCC is supported by other experienced climate scientists, all of whom are fully independent of the IPCC or any other official science organisation.

NIPCC produces summary reports of new scientific papers that contain data that do not favour the idea that dangerous, human-caused global warming is occurring. Many of these papers are either dismissed or ignored by the IPCC. For those who wish to consider all sides of the global warming issue, the weighty NIPCC reports form an essential counterbalance to those of the IPCC. In recognition of this, the 2009 and 2011 editions have recently been published in Chinese translation by the Scientific Information Center for Resources and Environment, Chinese Academy of Sciences.

All NIPCC publications can be downloaded free from the web site at *www.nipcccreport.org*.

Mr Gore's film *An Inconvenient Truth* — fact or fiction?
The film is a masterpiece of environmental evangelism.

Mr Al Gore's 2006 film, *An Inconvenient Truth*, comprises dramatic and beautiful scenes of imagined climate-related natural disasters. Collapsing ice sheets, shrinking mountain glaciers, giant storms, floods, searing deserts, ocean current and sea-level changes and drowning polar bears are all featured.

It is never explained that these events reflect the fact that we humans inhabit a dynamic planet, and that all the changes featured in the film have occurred naturally many times in the past — long before human activities could possibly have been their cause.

Troubled by the obvious propaganda intent of the film, in 2007 British parent and school governor Stewart Dimmock took exception to the UK government having provided all secondary schools in the UK with a video copy of *An Inconvenient Truth* for use in the classroom. He therefore sought an injunction from the High Court in London to direct the Secretary of State for Education to withdraw Mr Gore's film package from schools.

Though the Court declined to recall the film, in a famous victory for commonsense Justice Michael

THE UNSURE THING

NB) THE DOCUMENTARY "AN INCONVENIENT TRUTH" WILL BE INCLUDED IN THE AUSTRALIAN NATIONAL CURRICULUM DESPITE A FINDING BY THE U.K. HIGH COURT IN 2007 THAT THE FILM CONTAINS NINE ERRORS, CONCERNING SEA LEVEL RISES, PACIFIC ATOLLS, THE OCEAN CONVEYOR THREAT, CO₂ AND TEMP. GRAPHS, KILIMANJARO, POLAR BEARS, CORAL ETC

SPOONER

'Sure Thing' won the Melbourne Cup in 2010. October 2010.

Burton commented that 'the claimant substantially won this case', ruling that the science in the film was used 'to make a political statement and to support a political programme' and contained nine fundamental errors of fact. Justice Burton required that these errors be summarised in new guidance notes to be used as an accompaniment to future educational showings.

The nine errors identified in *An Inconvenient Truth* by Justice Burton consisted of erroneous or exaggerated statements about sea-level rise, evacuation of Pacific Islands, intensity of ocean current circulation, cause/effect relationship between increasing carbon dioxide and increasing temperature, melting glaciers on Mt. Kilimanjaro, the drying up of Lake Chad, and matters relating to Hurricane Katrina, polar bears and coral reefs. This comprises a list of most of the pin-up environmental scares that are used worldwide by proponents of dangerous anthropogenic global warming in support of their cause.

The London High Court judgement firmly typed Mr Gore, and the environmental organisations that work with him to spread global warming alarm, as proselytisers for an environmental cause and as abusers rather than users of scientific information.

Academies of science: statements from authority
Science is never based upon authority, however distinguished.

Given the complexities of climate science, it comes as no surprise that most public commentators base their opinions on statements of apparent authority rather than on an examination of the relevant scientific principles and supporting data. Most often, reference is made to the reports and policy statements of the IPCC, but advice from many of the world's major scientific academies, societies and government agencies is also relied upon. The citation of public statements by expert academies and organisations helps to convey a patina of independent assessment, but the reality is that such sources derive their material from the IPCC and then simply recycle it.

With the exceptions of the Polish and Russian Academies of Science, and significant scientific dissidents in China, most science organisations that have issued public statements on climate change support the notion that dangerous global warming is being caused by human greenhouse emissions. However, the statements that support warming alarmism are invariably issued under the imprimatur of a governing board or council, with no attempt made to canvas the views of the expert society membership. A recent and significant exception to this practice is the well-balanced new draft statement on climate change recently published by the Australian Geological Society[7]. This opinion was written only after the incumbent president, Professor Brad Pillans, had consulted widely with the society's membership as to their views, and thus ensured that the statement carries an imprimatur of genuine professional weight.

More generally, Thomas Kuhn, writing in *The Structure of Scientific Revolutions*, describes the processes whereby even firmly established scientific paradigms can eventually crumble under the weight of mounting contrary evidence, to be replaced by new paradigms. Thus argument from authority is the very antithesis of the scientific method, as reflected in the motto of the Royal Society of London which reads: 'Nullius in verba', meaning roughly 'trust the word of no one'.

Even were their views not subject to political or scientific fashion, that most of the world's academies of science favour a particular proposition is absolutely no guarantee that it is right. For example, during the mid-20th century the received scientific wisdom was that continental drift, as famously championed in 1912 by German meteorologist-geophysicist Alfred Wegener, was physically impossible. By 1970 new data had established that not only was continental drift possible but that it formed part of a set of fundamental physical mechanisms that now underpin the whole science of modern geology.

[7] *http://www.gsa.org.au/pdfdocuments/publications/TAG_165%20TAG.pdf,* p. 6)

Many other similar examples could be cited, and the implications for those who wish to place blind trust in 'expert' advice are obvious. As Nobel-prize winning physicist Richard Feynman once famously remarked, 'Science is the belief in the ignorance of experts'.

Given these facts, and that distinguished scientists can be found on both sides of the current argument about global warming, how is a government or a member of the public to decide the truth of the matter?

The answer has to be by undertaking a dispassionate, due diligence review of the scientific evidence, taking especial account of prevailing uncertainties. Though many citizens will consider that the IPCC was set up to discharge just such a function, the reality is that it was not. Rather, the IPCC operates at two levels: first, by Working Group I undertaking a fundamental scientific assessment of the published literature regarding emissions-related climate science; and, second, the preparation of social advocacy policy by Working Groups II and III which emphasises a political agenda. The twin roles of objective scientific assessment and political policy analysis inevitably conflict with one another, which makes the IPCC an inadequate organisation to be treated as an independent science auditor.

It is a remarkable fact that, despite many requests, no official IPCC-independent review of the evidence for and against dangerous warming has ever been undertaken in a western country. The closest to a document of this type in Australia was the 2009 audit of the IPCC's advice to the Department of Climate Change that was performed by four senior scientists at the request of Senator Fielding.[8] Amazingly, however, many lobby groups and political organisations work assiduously to prevent just such reviews from happening. Readers who think through the implications of these facts will be able to draw their own conclusions.

What is the Climate Commission?

Government appointees who provide climate information in support of government policies.

In accordance with a commitment given during the 2010 federal election campaign, the Climate Commission was established by Prime Minister Julia Gillard's Labor government in February, 2011, and operates from within the federal

---

[8] See *http://www.quadrant.org.au/blogs/doomed-planet/2011/04/due-diligence-reports*

Department of Climate Change. The Commission was created with 4-year funding of $5.6 million, and a brief to provide all Australians with independent and reliable information about (i) the science of climate change, (ii) the international action that is being taken to reduce greenhouse gas emissions, and (iii) the economics of a carbon price. The Commission conducts public information meetings around Australia, and has published a number of reports about global warming science and policy.

The commissioners in 2012 were Tim Flannery (chair, scientist), Roger Beale (economist), Gerry Hueston (businessman), Lesley Hughes (ecologist), Veen Sahajwalla (materials engineer) and Will Steffen (scientist). The Commission also takes advice from a panel of eight leading Australian contributors to the IPCC assessment reports.

The material provided in the Commission's public meetings and reports adheres closely to the IPCC's policymaker summaries of global warming, supplemented by local material provided by CSIRO and the Bureau of Meteorology. Several of the reports have been severely criticised by independent senior Australian scientists (see footnote 8, p. 49). While these criticisms remain unrebutted (as is also true for many similar criticisms that have been made of the IPCC's conclusions), it is hard to view the advice of the Commissioners as independent, authoritative or reliable.

What is the hockey-stick and why was it important?
A temperature reconstruction contrived to represent alarming 20th century warming.

In 1998, scientists Michael Mann, Raymond Bradley and Malcolm Hughes (MBH) published a graph of reconstructed temperature between AD 1200 and 1900 in the prestigious *Nature* magazine.
The graph was based upon analysis of 183 tree-ring and a few ice core records from sites across the northern hemisphere, merged at its younger end with the the measured thermometer temperature record from 1900 to 2000 AD. Overall, the graph (Fig. 8, p.51) exhibits a gently declining overall temperature up to 1900, followed by a sharp rise in temperature thereafter. The graph has the general shape of an ice-hockey stick, and hence the name.

Variations on the original graph were published in 1999 and 2000, with additional tree ring data that extended the record back to 1000 AD. The extended graph was included in the IPCC's 2001 Third Assessment Report,

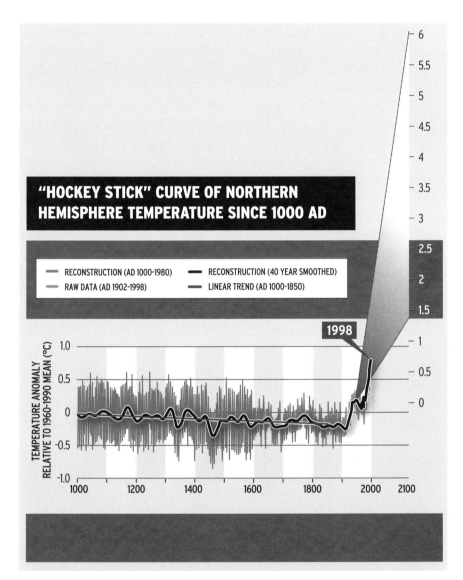

Fig. 8. Typical public depiction of the Mann, Bradley & Hughes hockey-stick curve of reconstructed northern hemisphere temperature since 1000 AD (PAGES, 2001). The graph comprises the conjunction of two separate temperature records (tree ring statistics up to 1900 AD, and thermometer measurements thereafter), with an added and speculative computer projection beyond 2000 AD after the IPCC. Both the Medieval Warm Period and Little Ice Age are absent (compare Fig. 5, p.29). The red colour bars serve to convey an impression of alarming warming.

and played a central role at the report's launch. The central message from the graph was that Earth's climate was stable prior to the rapid expansion of industrialisation during the 20th century, the latter being accompanied by rapid global warming.

The graph appeared many times in the 2001 report. One such appearance was accompanied by the comment that 'the 1990s has been the warmest decade and 1998 the warmest year of the millennium for the northern hemisphere': a message that was subsequently marketed relentlessly by proponents of dangerous AGW. The MBH hockey stick was thus of paramount importance to the IPCC assertion that dangerous human-caused climate change had started to occur in the 20th century.

But all was not as it seemed, for the hockey stick curve contradicted well established earlier scientific understanding in two conspicuous ways. First, it showed that temperatures held mainly steady, though with a slight long-term decline, between AD 1000 and 1900; well documented climatic episodes such as the Medieval Warm Period and the multi-troughed Little Ice Age did not appear, despite the fact that many individual high resolution climate records depict these episodes with clarity (compare Fig. 5, p.29). Second, the rapidly ascending 20th century blade of the curve appears to show temperature increasing at an unprecedented rate and to an unprecedented magnitude by the end of the century, yet high quality speleothem and tree ring records from the southern hemisphere showed not a trace of the dramatic warming that was now being claimed for the northern hemisphere.

Despite its many inadequacies, the MBH graph gave birth to the legend that late 20th century temperature was increasing to an unprecedented peak at a dangerous rate. The alarm was fanned even further by versions of the hockey stick diagram like Fig. 8 (p.51), with steeply rising computer model temperature projections tacked on to its top end.

Noticing the conflict between the MBH hockey stick and other more direct climate records, many scientists were suspicious of these results. Soon new studies of temperature change over the last 500–2000 years began to be published that reaffirmed the traditional interpretations of recent climatic history. Two Canadian experts in statistics, mining analyst Steve McIntyre and economist Ross McKitrick, attempted to replicate the initial MBH hockey stick graph. The difficulty in replicating the work lay not only in the unusual statistical analysis involved, but also in persuading the authors and publishers of the original paper to release the data on which the research was based.

In time, though, enough original data was released for the hockey stick curve to be replicated, and, in a pair of forensic science-audit papers, McIntyre and McKitrick showed that the MBH hockey stick graph was based upon gravely flawed statistical analysis. The pièce de résistance was the demonstration that the statistical analysis used by MBH (technically, a principal components analysis) had a strong intrinsic tendency to produce 'hockey stick' shaped output curves even when fed with random data sets.

Not surprisingly, a public furore arose when these results were published, but in time other investigators vindicated McIntyre and McKitrick's work, including not least in an independent statistical audit performed by senior statistician Professor Edward Wegman for a U.S. congressional committee.

Perhaps inevitably, the MBH affair has become one of the most celebrated misadventures of modern science, as described in close detail in Andrew Montford's best-selling book, *The Hockey Stick Illusion: Climategate and the Corruption of Science*.

What was Climategate and why was it important?
Scientists behaving badly.

On October 12, 2009, the climate correspondent for BBC's *Look North* programme in Yorkshire and Lincolnshire, Paul Holmes, received an email in response to a 9 October article that he had written entitled 'What has happened to global warming?' Attached to the email were copies of several thousand emails that had been interchanged between Director Phil Jones and his staff at the University of East Anglia's Climate  Research Unit (CRU) and their colleagues elsewhere.

CRU's established reputation is based on its work in collecting historical meteorological records and constructing a land-based global temperature history. These land-based data were combined with an independent record of ocean surface temperatures gathered by the UK Meteorological Office's Hadley Research Centre to establish the global near-surface temperature record that is used by the IPCC (Fig. 1, p.17).

Despite its high public significance, the BBC chose not to report the scoop that it had received. Knowledge of the existence of the emails, and their contents only became public when copies were received and commented on by three U.S. climate blog sites on November 19 — The Air Vent, Climate Audit and Watts Up With That. Subsequently, the BBC carried the story on its news

GOLDILOCKS WAS NEVER THE SAME AFTER SHE READ THOSE HORRIBLE EMAILS

The release of the 'Climategate' emails had a devastating effect on public
confidence in the Rudd government's Carbon Pollution Reduction Scheme. May 2010.

channels, but only in a reactive way that had obviously been provoked by the
blog postings. What, then, did the leaked package of CRU papers contain, and
why was their release so significant?

The papers comprised more than 60 Mb of emails, reports and computer
code relating to CRU's research activities. The emails, which have been published on several websites and in printed format, revealed in some detail the
operation of an international network of scientists who were advocates of
dangerous AGW and closely linked to the IPCC. The emails revealed how
these scientists were constantly strategising on how to get their advocacy
message out to the public, and, equally, how they could obstruct the attempts
of independent scientists to have the issue of global warming discussed in a
balanced and rational manner.

The most disturbing material in the package of documents, however, was
in several reports that contained computer coding. This code, and accompanying programmers' notes, showed unequivocal evidence for data manipula-
tion. And public exposure Climategate duly got. No fewer than four separate
investigations were convened to review the Climategate matter, including
one in the U.K. parliament, as IPCC-supporters in high places desperately
tried to whitewash the incident into innocence. Unfortunately these enqui-

ries were largely politically motivated and did not address the influence of the network participants on the independence of the IPCC process. Although no criminality was identified, some individuals were admonished on procedural grounds and overall the enquiries entirely failed to remove a dubious smell that still lingers.

As well, the Climategate enquiries were too little and too late, for the hare of genuine investigative enquiry was running freely in the world media long before the reviews were completed. Public clamour has continued to mount since 2010, as more and more discoveries have been made that reveal the partiality of the IPCC. We now have knowledge of not just Climategate, but also of Glaciergate, Amazongate, NASAgate, Pachaurigate and what one blog has listed as more than 100 similar IPCC-related biases.

The Climategate incident separates two different worlds and marks a paradigm shift. Up to 2009, IPCC authority was mostly accepted by scientists and politicians alike as the arbiter for all matters to do with anthropogenic global warming. The post-Climategate world, from 2010 onward, is one in which the IPCC has lost most of its once-high credibility except among committed advocates for dangerous AGW.

The ad hominem zoo: sceptics, deniers, agnostics and warmaholics
Perjorative name-calling is the stuff of politics, not science.

It is the mark of a scientific debate that participants confine their discussion to unemotional examination of the factual and theoretical material in hand.

How can it be, then, that virtually every public discussion of the global warming problem involves the use of perjorative, ad hominem terms such as climate sceptic, climate denier, contrarian, rejectionist, obscurantist, confusionist or flat-Earther; and in description of those who hold the opposing point of view — climate alarmist, climatist or warmaholic. These terms range from descriptive (alarmist) through amusing (warmaholic) and silly (sceptic) to deliberately offensive (denier)

To term a person a climate sceptic is simply to reinforce that he or she is a scientist. For all good scientists, including both those who advise the IPCC and the NIPCC, are sceptics; it is their professional job to assess evidence as it relates to particular hypotheses. To not be a sceptic of a hypothesis that you are testing is the rudest of scientific errors, because it means that you are committed to a particular outcome: that's faith, not science.

Most people termed climate 'sceptics' or 'deniers' by their opponents, and all true scientists in general, are in fact climate 'agnostics'. This is to say that, in advance of analysis, they have no particular axe to grind regarding the magnitude of the human influence on global climate. Rather, they just want the facts to be established, and for the interpretations to then fall where they most logically lie.

The reason why public discussion of the global warming problem so often involves the widespread use of ad hominem descriptors is because of the absence of open discussion on this highly contentious public issue. Those advocating dangerous anthropogenic global warming prefer the shield of 'expert' and 'authoritative' statements to logical defence of their scientific tenets; they also resort to denigration to categorise those with whom they disagree. The media, ever keen to introduce economy into their stories, love to label the people whose views they report; should such a label carry emotional or political baggage, then so much the better, for that provokes human interest.

It goes without saying that professional people who contribute to the public debate on global warming should avoid using perjorative labels to describe those who have different views. In addition, and as a rule of thumb, any piece of writing or broadcast that sprinkles these terms around like confetti, as (regrettably) most media reports do, can straightaway be flagged as likely to be inaccurate or biased.

Finally, the introduction of terms such as 'denier' or 'denialist' into the public climate debate, with their connotations of holocaust denial, is a deliberate strategy to move debate from scientific substance to political emotion. The ploy is to label those who support the dangerous AGW hypothesis as ethically superior to those who challenge the claim on scientifically rigorous grounds, and it serves only to cheapen those who practice it.

Does he who pays the piper call the tune?
In matters of science and ethics, never.

A related technique to the ad hominem labelling of a person with whose views you disagree is the custom of querying their sources of financial support. The implication is that if a funding source can be badged as 'unsavoury' then any science that results will by definition be tainted and unreliable.

Commonly, for example, after someone has been labelled as a climate 'sceptic'

they are then linked to an organisation, no matter how tenuously, that
can be accused of accepting money
from coal or petroleum companies.
The latter are badged as Big Coal
and Big Oil and are seen to represent irredeemably vested interests,
hell-bent on 'polluting' the atmosphere. It is apparently lost on the
accusers that whenever they turn
on an electric switch, or fill their
car with petrol, they are supporting
the need for the existence of Big Coal and Big Oil.

It is common practice across business and politics to hire professional advocates for a cause, just as an advocate is hired for a legal defence. Advocacy
in itself can be a noble endeavour, and a scientist or engineer is justified in
using their expertise and skills to advocate for a particular cause. However, an
ethical boundary is crossed should an advocate manipulate data and information to self-interestedly misrepresent the real situation. Scientists and engineers, whether in public or private employment, are respected as professionals
by the public according to the degree that they deploy their expertise for the
ultimate general good.

In contrast to business and politics, science operates in an intentionally value-
free fashion by erecting ideas (hypotheses) about the world around us, gathering
appropriate experimental or observational data, and then testing the original
idea against the available data. It can be argued by sociologists of post-modernist
leaning that science, like all other human activities, is a social construct. Such
views overlook the fact that scientific hypotheses are either validated with time,
or are amended or discarded in the light of new evidence.

Who might have funded a given piece of scientific research is simply irrelevant. Whether Mother Teresa or Genghis Khan put in the cash, an output,
being scientific, is testable against the facts. Testing hypotheses against facts is
what scientists DO — and according to a set of strict logical rules.

The validity of a scientific truth does not depend upon the character of
those who might have accomplished the research, or subsequently come to
accept it, these things being just as irrelevant as who might have provided the
funding. Whether you like the scientist, or her mother, or his politics, religion
or funding agency has absolutely nothing to do with whether a piece of scientific research is valid. It may be hard to believe in a post-modern world, but

in the progress of science who paid for the data to be gathered and assessed, and who performed the research, are both equally irrelevant. Were that not to be the case then any study in question would simply not be a scientific one.

Use of ad hominem criticism and funding-smear arguments are therefore both intellectually dishonest. Tellingly, such techniques are invariably used as a substitute for participating in a discussion of the scientific matters under consideration. In the global warming game, that is for the very good reason that genuinely open and balanced discussions of the relevant science invariably lead to the conclusion that dangerous human-caused warming is neither ocurring now nor an imminent threat.

How accountable are non-governmental organisations?
They are accountable to no one.

NGOs hold a particular status in society. Each has its own agenda and, being classified as not for profit, an NGO is generally regarded as having altruistic motives. In size they range from small community-based collectives to multi-national behemoths. Financial support for NGOs comes from individual donations, business donations, commercial activities and allied government programmes.

As a class, environmental NGOs play a major and constructive role in society. In many cases they are no more than a community-based group addressing a specific or more general local need. This may be a small creek or river reserve that does not receive government or council attention, but which is considered a local asset for recreation, or as a local natural habitat. However, at the other end of the scale, a number of large national and international environmental NGOs aim to achieve their specific, far-reaching and political objectives through community advocacy, citizen activism and high level lobbying.

The World Wildlife Fund (WWF) and Greenpeace were founded in 1961 and 1972, respectively. Today, they are the Big Two in in an environmental industry of great power and influence that now comprises literally thousands of lobby groups of all shapes and sizes. For example, at the COP-17 climate conference in Durban in 2011, the 5,884 (40%) registered NGO participants present all but matched the 6,172 (42%) government representatives attending. Commanding cash flows of billions of dollars, and with global spread, the largest environmental NGOs have come to be more powerful than many sovereign governments. Indeed, at some UN meetings it is not unusu[al] smaller countries formally represented by NGO staff.

As the environmental movement grew in size through the 1970[s,] interest began to be expressed in the influence that human activity might be having on the climate. In addition to concerns over the deleterious impact of power station particulate and sulphate emissions, interest centred on whether industrial carbon dioxide emissions might cause dangerous global warming. Given that carbon dioxide is undeniably a greenhouse gas, and that humans were indeed adding extra to the atmosphere (albeit small amounts in terms of natural flows), this was an entirely sensible question to raise at the time. About ten years later, in 1988, the IPCC was formed to address exactly this question, and 24 years on again it remains unresolved.

Those 24 years have seen thousands of scientists expend well over $100 billion in studying the influence that human-related emissions may be having on climate. Given these intensive efforts, the absence of a measurable or un-equivocal human imprint in the recent temperature record and the absence of any global warming trend at all over the last 16 years both point to frailty in the dangerous AGW hypothesis. A reasonable default conclusion is that any human influence on the global climate lies within the noise of natural variability.

A similar conclusion has been reached recently by a group of 21 leading IPCC-linked scientists (Ben Santer and colleagues) in a computer modelling paper published in 2012 in the Proceedings of the US National Academy of Sciences.[9] From their modelling, these scientists have shown that any human warming influence on global climate must lie within the natural variability of climate. It is therefore by definition not dangerous in the short term, and probably not in the longer term either.

Unfortunately, however, the environmental movement, and most espe-cially its big players, are much better at shouting than they are at listening, and they have not heard this message. Or, perhaps more likely, they have heard the message but are determined to subjugate or ignore it. But whatever the reason, accountable to no one and with contempt for empirical science, major NGOs and a cast of smaller environmental groups today continue to evangelise for industrial countries to reduce their carbon dioxide emissions, irrespective of the economic and social costs involved. Worse, the costs of misguided attempts to cut carbon dioxide emissions are visited most severely

[9.] Santer, B.D. et al. 2012. Identifying human influences on atmospheric temperature. Proceedings U.S. National Academy of Sciences, Nov. 29. *http://www.pnas.org/content/early/2012/11/28/1210514109.abstract.*

on the poorest members of western countries, and, preferentially again, on the citizens of undeveloped countries.

It is not for nothing that the former president of the Czech Republic Vaclav Klaus has concluded that radical environmentalism is a greater threat to modern society than communism ever was; and as a professor of economics under a communist government, and then a minister for finance in the new Czech government that replaced communist rule in 1989, he should surely know. The agenda of modern radical environmentalism is not saving or improving the environment, if ever it was, but to control and fundamentally restructure capitalist societies through the manipulative management of energy resources.

What funds are available in support of global warming studies?

Lots: but the overwhelming majority of funds flow to those who profess to believe in dangerous AGW.

It is often stated that action to restrict anthropogenic greenhouse gas emissions and prevent dangerous global warming is being hampered by the actions of well-funded skeptics and deniers in the pay of Big Oil and Big Coal (see above: Does he who pays the piper call the tune?). It is implied that these so-called 'dirty industries' are swaying politicians to prevent necessary action to save the planet from dangerous climate change. Without wishing to give credence to such an irrational style of argument, it is salutary nonetheless to compare the funding outlaid for research and advocacy in support of the dangerous AGW hypothesis with that raised by those challenging the need for alarm.

It is difficult to get accurate figures that answer this question. However, Joanne Nova has provided the following selected figures on her blog (Table 2), and they illustrate the scale of the problem well enough.

The arbitrary sample of 2009/10 expenditure on global warming-related research that Joanne provides, drawn from publically available figures, sums to $2.5 billion for US government sources and $2.6 billion for environmental NGOs plus public research groups. All of this money is provided for activity that is anchored to the IPCC's rationale that human greenhouse gas emissions are causing dangerous warming. Meanwhile, on the other side, arguing the case for an independent assessment of global warming hazard, sits the Heartland Institute – a US libertarian think-tank that had a total budget in 2009 of $6.4 million, of which only $390,000 was spent directly on climate-related issues.

Entity	USD	
Greenpeace	$300m	2010 Annual Report
WWF	$700m (€524m)	2010 Annual Report
Pew Charitable Trust	$360m	2010 Annual Report
Sierra Club	$56m	2010 Annual Report
NSW climate change fund (just one random govt. example)	$750m (A$700m)	NSW government
UK university climate fund (just another random govt. example)	$360m (£234m)	UK government
Heartland Institute	$7m	(actually $6.4m)
US government funding for climate science & technology	$7,000m	Climate Money, 2009
US government funding for 'climate related appropriations'	$1,300m	USAID, 2010
Annual turnover in global carbon markets	$120,000m	2010 Point Carbon
Annual investment in renewal energy	$243,000m	2010 BNEF
US govt funding for sceptical scientists	$0	

Table 2. Estimated climate change funding expended by or available to a selection of major governments and NGOs. (after Joanne Nova, at *http://joannenova.com.au/*)

Even treating the Heartland Institute's budget as entirely spent on climate change issues, that represents an astonishing ratio of money for expenditure on the pro-dangerous AGW and anti-dangerous AGW cases of about 800:1. Further, the limited sample of organisations and countries listed in the table means that the worldwide expenditure is very much greater overall, in total probably at least twice as great as the amounts listed in the table and even more biased towards dangerous warming alarmism.

Effectively, the climate change debate amounts to putting modern industrial societies on trial for crimes against the climate in a kangaroo court of peer-reviewed science; developing countries are innocent, though denied development. There are virtually no defence counsel to be seen, nor any money to fund them.

Since the demise of the Australian Science and Engineering Council (ASTEC) in 1999, Australian scientific issues, and especially those that relate

Bob Carter was unfairly accused of impropriety for accepting consultancy fees from the Heartland Institute. Prime Minister Julia Gillard and climate commissioner Prof. Tim Flannery. February 2012.

to contentious environmental matters, have lacked any mechanism of expert independent audit, similar to that which the Productivity Commission is intended to provide for economic issues. Given the social and economic disruption that is now often caused by lobby-group driven, quasi-environmental issues, this deficiency urgently needs rectifying, for example by the creation of a Science Audit Commission. The job of any such commission would be to ensure that government be informed of both the prosecution AND defence case before expending money on the prevention of alleged environmental crimes. In other words, the contestability of expert evidence needs to be formally required as part of any process of government policy formulation.

A similar proposal received much discussion and serious consideration in the late 1960s in the USA, under the intended name of a Science Court. It is not, however, the name that matters, but that any such body be staffed by independent scientists who are charged with the duty of ensuring that the government, and taxpayers, are not blindsided by scientific bias, exaggeration or error.

Why all this talk about carbon instead of carbon dioxide?
Because it evokes the image of dirty smokestacks, long a thing of the past in Australia.

Since the 1970s, green lobby groups have come to realise the power that resides in defining the language used in environmental debates — generally by substituting deliberately emotionally-charged words for factually accurate ones.

Every time a politician or public figure uses the word 'carbon' when they mean 'carbon dioxide' they signal either their ignorance of the difference between an element and a molecule, or an intent to deceive.

Carbon dioxide is a colourless, odourless gas that occurs naturally in the atmosphere. Moreover, it is vital for life on Earth. Carbon dioxide from the atmosphere is taken up and incorporated into vegetation during growth through the process of photosynthesis; carbon dioxide is returned to the atmosphere during respiration and decay of dead biomass. Carbon dioxide is not a pollutant and more carbon dioxide in the atmosphere is beneficial to plant growth (see IV: Is atmospheric carbon dioxide a pollutant?)

Reference to 'carbon' and 'carbon pollution' in the context of carbon dioxide emissions is a deliberate attempt to link the unburned residue of combustion, the smoke and noxious gases of 19th and early 20th century smokestacks, with carbon dioxide. But in western countries, clean air laws now ensure that dirty smokestacks of power generators and industry are a thing of the past. Technological innovation has ensured the demise of such 'dirty' industry by-products, and today the genuine pollutants of soot and ash are filtered out at source, together with noxious sulphurous and nitrous oxides.

The emissions from modern smoke stacks are largely composed of two non-polluting, environmentally beneficial gases, to wit water vapour and carbon dioxide. This therefore marks as duplicitous the constant misuse of pictures of power station chimneys emitting steam — just like a boiling kettle, and no more alarming — as a backdrop to news and current affairs stories about the greenhouse warming issue.

To term environmentally beneficial carbon dioxide emissions as (dirty implied) 'carbon' or 'carbon pollution' is an abuse of logic, an abuse of language and an abuse of science.

Is the science really settled?

No. Scientific knowledge is always a moving feast.

A common argument advanced by those urging the reduction of carbon dioxide emissions is that the science of the matter has been thoroughly investigated (which in some parts it has) and is 'settled' (which is simply untrue).

Science is about the collection of facts and experimental data in pursuit of testing hypotheses; new facts very often lead to modification or rejection of a previously established working hypothesis. The history of science is replete with examples of how accepted wisdom has been overturned as new technologies have provided previously unavailable data that is in conflict or provides new insights. The climate system is very complex and the science is continually evolving. Thus in contrast to policy, science is never 'settled'.

The current uncertainty of knowledge in climate science is well summarised by the following statement, which is extracted from a December 8, 2009 letter addressed to the Secretary-General of the United Nations. The letter was signed by 166 independent, well qualified scientists.

Climate change science is in a period of 'negative discovery' — the more we learn about this exceptionally complex and rapidly evolving field the more we realise how little we know. Truly, the science is NOT settled.

Therefore, there is no sound reason to impose expensive and restrictive public policy decisions on the peoples of the world without first providing convincing evidence that human activities are causing dangerous climate change beyond that resulting from natural causes.

Before any precipitate action is taken, we must have solid observational data demonstrating that recent changes in climate differ substantially from changes observed in the past and are well in excess of normal variations caused by solar cycles, ocean currents, weather cycles (El Nino, etc.), changes in the Earth's orbital parameters and other natural phenomena.

The argument that 'the science is settled' is a ploy aimed at shutting down public discussion on an issue that in reality remains deeply uncertain and highly contentious. Use of the argument is therefore a sure sign of a political agenda, through which those advocating political action prevent the frail scientific evidence for dangerous warming from being scrutinised. The widespread use of this technique by scientists, politicians and media commentators who view AGW as a crisis represents a clumsy (though regrettably often effective) attempt to negate sensible public discussion about global warming; and especially to negate the balancing contributions that independent expert scientists can bring to the table.

But don't 97% of all scientists say that dangerous warming is occurring?

No. A majority of scientists have expressed public scepticism about dangerous warming.

The assertion that 97% of all scientists agree that dangerous global warming is occurring is a fantasy.

First, because no one knows, or can know — obviously — what 'all scientists' think.

Second, because the two studies that most journalists refer to when they repeat these fantastical figures are deeply flawed.

These studies are:

- An April 2008 poll conducted by Professor Peter Doran and masters student Margaret R. K. Zimmerman at the University of Illinois, Chicago. The survey results were summarised in a paper published in January 2009 in the science journal EOS.
- A 2010 paper in the Proceedings of the National Academy of Sciences of the United States (PNAS) by William Anderegg and co-authors.

The Doran and Zimmerman study, although it polled expert scientists (90% from USA; so much for worldwide), has been thoroughly debunked by many writers, and there is little point in repeating their detailed criticisms here. The poll's credibility as a reliable measure of the stance of climate scientists on dangerous AGW is completely undermined by the fact that the sample size was tiny (77 persons), and that many respondents themselves commented unfavourably on the design of the poll questions and other procedural defects.

Moving to the Anderegg study, contrary to popular belief, it did not poll expert scientists at all. Instead, the authors of the paper simply evaluated the publication record of selected scientists whom they considered representative of the global warming debate. They did this by counting the number of articles published in academic journals by 908 climate researchers (defined as people who have published 'a minimum of 20 climate publications'). The study found that 97–98% of the most prolific 200 climate researchers, so defined, appeared to believe that 'anthropogenic greenhouse gases have been responsible for most of the unequivocal' warming of the Earth's average global temperature over the second half of the 20th century'.

The Anderegg survey suffered from major biases, including: publication bias (papers that claim to provide evidence for warming are much more likely to get published); joint-authorship bias (the apparent status of many 'leading' climate scientists being boosted by the practice of including large numbers of author 'mates' on papers); age bias (many of the most knowledgeable climate agnostic scientists are now retired, no longer actively publishing and therefore were not considered in the study); and editorial bias (it being well known since the Climategate affair that a small clique of influential government scientists constantly works behind the scenes to get academic journal editors to reject sceptical articles).

In essence, then, neither the Doran and Zimmerman nor the Anderegg study provide rigorous scholarly results that bear on scientific opinion about dangerous AGW.

Finally, there is a third reason why the general belief that most scientists believe in dangerous AGW is fanciful. It is that many thousands of independent scientists — all of whom acknowledge that carbon dioxide is a greenhouse gas and that human activities are contributing 'extra' carbon dioxide to the atmosphere — have signed statements to the effect that no evidence exists that dangerous global warming is occurring as a result. A typical statement, signed by 166 expert scientists and currently posted on the website for the International Climate Science Coalition, reads:

> *We, the undersigned, having assessed the relevant scientific evidence, do not find convincing support for the hypothesis that human emissions of carbon dioxide are causing, or will in the foreseeable future cause, dangerous global warming.*

Table 3 summarises some of the many other public statements signed since 1992 that espouse a similar view.

Given the fact that just 53 scientists appear as named authors of the critical Chapter 9 of the fourth and most recent IPCC report (Understanding and Attributing Climate Change), the majority scientific opinion clearly favours that dangerous AGW is not currently a problem that requires urgent political action — faux 97% opinion polls to the contrary notwithstanding.

1992 **The Heidelberg Appeal (Rio Earth Summit).**
Signed by more than 4,000 scientists (including 70 Nobel laureates, from 106 countries.
http://www.sepp.org/policy%20declarations/heidelberg_appeal.html

1997. **The Leipzig Declaration (Kyoto Protocol).**
More than 100 signatories, including editors of Climate Research and Atmospheric
Research, former president of the U.S. National Academy of Sciences.
http://www.sepp.org/policy%20declarations/home.html

1998 (2007) **The Oregon Petition (IPCC 3AR).**
Signed by more than 31,000 professional persons, of whom 9,029 have PhD
degrees, including 2,660 physicists, geophysicists, climatologists, meteorologists,
oceanographers and environmental scientists, and 5,017 scientists from other
disciplines. Includes the statement: *Not only has the global warming hypothesis failed the
experimental test, it is theoretically flawed as well.*
http://www.oism.org/pproject/

2007 **IPCC conference (Bali).**
Open letter to UN Sec.-Gen., with 103 signatories, 23 of them Emeritus Professors.
http://climatescienceinternational.org/index.php?option=com_content&view=article&id=445

2008 **Manhattan Declaration (Heartland Conference).**
Endorsed by 1,500 individuals, half of whom are well trained in science and/
or technology, and over 200 specialise in climate research.
*http://climatescienceinternational.org/index.php?option=com_content&task=view&id=
37&Itemid=54.*

2009 **IPCC Conference (Copenhagen).**
Open letter to UN Sec.-Gen. from 200 climate experts during the
Copenhagen Climate Conference.
http://www.copenhagenclimatechallenge.org/

2010 **IPCC Conference (Cancun).**
More than 1000 International Scientists Dissent Over Man-Made Global
Warming Claims.
*http://www.climatedepot.com/a/9035/SPECIAL-REPORT-More-Than-1000-
International-Scientists-Dissent-Over-ManMade-Global-Warming-Claims--Challenge-
UN-IPCC--Gore*

2012 **COP-18 Conference (DOHA).**
Open letter to the UN Sec.-Gen. from 132 expert scientists and social
scientists during the UN FCCC meeting in Qatar. *We … ask that you
acknowledge that policy actions by the UN, or by the signatory nations to the
UNFCCC, that aim to reduce CO_2 emissions are unlikely to exercise any significant
influence on future climate. Climate policies therefore need to focus on preparation for,
and adaptation to, all dangerous climatic events however caused.*
http://www.climatescienceinternational.org/index.php?option=com_content&view=article&id=761

Table 3. Listing of some major statements signed by independent scientists since 1992 that
express caution and scepticism about the hypothesis of dangerous global warming caused by
human-related carbon dioxide emissions.

But isn't there supposed to be a consensus about global warming?

'If it's consensus, it isn't science. If it's science, it isn't consensus. Period.'

The headline quotation above comes from a famous critical essay written in 2003 by the late Michael Crichton. Though he trained as a medical doctor, Crichton's writings invariably display a firm and exceptionally clear-sighted grasp of the methodology and philosophy of science itself. In the article, Crichton wrote:

> *I regard consensus science as an extremely pernicious development that ought to be stopped cold in its tracks. Historically, the claim of consensus has been the first refuge of scoundrels; it is a way to avoid debate by claiming that the matter is already settled. Whenever you hear the consensus of scientists agrees on something or other, reach for your wallet, because you're being had.*

As you consider Crichton's view, it is worth reflecting as to when you last heard a scientist say 'there is a consensus that the Sun will rise tomorrow'. The answer, of course, is never. Instead, scientists make the confident statement that 'the Sun will rise tomorrow', based upon repeated empirical testing (which we all participate in every day) and on the analytical understanding conferred by Copernican and Newtonian theory.

It follows that the very use of the phrase 'consensus' in science tells you that a matter under consideration is not settled. Therefore, statements such as 'there is a consensus that dangerous global warming will occur' betray socio-political rather than scientific intent.

There is no Law of Climate, and the available empirical data shows that global temperature has varied through time, often for reasons that are not understood. That 20th century warming in part ran parallel to rising industrial carbon dioxide emissions and atmospheric concentration (Fig. 7, p.36) may simply be coincidental. Claims that 'there is a consensus that dangerous AGW is occurring' are more a reflection of a political agenda than a reflection of scientific knowledge.

Is there any common-ground amongst scientists who argue about this matter?

Yes, lots, but unfortunately the press and politicians have a very poor understanding of this.

Though you wouldn't know it from the antagonistic nature of public discussions about global warming, a large measure of scientific agreement and shared interpretation exists amongst nearly all scientists who consider the issue. The common ground includes:

- that climate has always changed and always will,
- that carbon dioxide is a greenhouse gas and warms the lower atmosphere,
- that carbon dioxide emissions are accumulating in the atmosphere as a result of industrial activity,
- that a global warming of between 0.4° and 0.7°C occurred in the 20th century, and
- that global warming has been in hiatus over the last 16 years.

The scientific argument over dangerous AGW is therefore about none of these things. Rather, it is almost entirely about three other, albeit related, issues:

- the amount of net warming that is, or will be, produced by human-related emissions,
- whether any actual evidence exists for measurable human-caused warming over the last 50 years, and
- whether the IPCC's computer models can provide accurate climate predictions 100 years or more into the future.

These issues are described and discussed in more detail in the answers given to questions subsequent to this one, especially those in sections III-VII.

What was the Kyoto Protocol?
An expensive and unsuccessful attempt to cut global carbon dioxide emissions.

The Kyoto Protocol was an international United Nations agreement that was aimed at limiting carbon dioxide emissions to 5% below 1990 levels. The base year was chosen carefully to advantage European nations, whose politicians knew they would benefit from the closure of many uneconomic greenhouse gas-producing industries after the fall of the Berlin Wall and liberation of eastern Europe. Thus politics, not science or environmental benefit, has been at the heart of all Kyoto discussion and activity ever since.

The protocol was an initiative that emerged from the 1992 Rio Earth Summit. It was based on the assertion that human-related carbon dioxide

emissions would cause dangerous AGW. It was accepted that rich, western nations, which have already benefited from industrialisation, should reduce their emissions first, with developing nations to join the scheme later. Those lobbying for the Kyoto Protocol took great heart from the successful completion of the 1987 Montreal Protocol to limit CFC emissions, an attempt to prevent damage to the ozone layer that was viewed as a model to follow.

The Kyoto Protocol was adopted in 1997, but required ratification from member countries during the period up to 2005 when it came into force. In the event, 37 industrialised nations signed up to reducing their emissions between 2008 and 2012, but that did not include USA (nor, initially, Australia) or any of the large developing countries like China, India and Brazil — these and other non-signatories together contributing 85% of global emissions.

Even if all countries that had signed the protocol had met their obligations (which was never going to happen: Canada, New Zealand, Russia and Japan formally withdrawing, in Canada and Japan's cases to avoid the embarrassment of failing to meet their targets), the theoretical effect on world temperature was unmeasurable. One widely accepted model projected that the amount of warming prevented by 2100 if all signatory nations were compliant would be only 0.02° C.

The Kyoto Protocol terminated on December 31, 2012, and has proved to be an expensive lame duck. Applying to countries that only produce 15% of world emissions, the protocol lapsed leaving the world with 58% more greenhouse gases than in 1990 compared with the 5% reduction sought by its proponents.

At the Doha COP-18 climate conference in December, 2012, 195 countries agreed to pretend that the protocol will live on, by 'extending' it pending ratification of a new treaty by 2015 that would take effect in 2020. This face-saving announcement probably represents the death rattle of the Kyoto Protocol.

As the *Wall Street Journal* recently put it: 'Count this as another eco-cure that arrived with a bang and departed, as so many of them do, with a whimper'.

III

THE RECORD OF
CLIMATE CHANGE

How do we know about ancient climate?

From Earth's geological history as preserved in sedimentary rocks.

A major part of geological study is concerned with unravelling the Earth's past, a deep time environmental record based upon information contained in ancient sediments and sedimentary rocks like mudstone, sandstone and limestone. Australia has amongst the oldest sedimentary rocks in the world, which date from approximately 3.5 billion years ago, a period called the Archaean. Geologists have assembled an environmental and climatic record based on the study of these and younger sedimentary rocks, though the older parts of this record are understandably fragmentary. The reconstructed record suggests that for about the last 2 billion years the Earth has had an atmosphere and oceans similar in physical and chemical properties to (though not identical with) modern counterparts, with organised multi-cellular life also present for the last 600 million years.

Deep time climatic studies provide important context, but the most critical geological evidence against which to assess modern climate change comes from sedimentary rocks less than about 10 million years old. Vital archives of information have been extracted from long sediment cores beneath the ocean seabed (back to 10 million years and beyond) and ice cores through the Antarctic (back to 1 million years) and Greenland ice caps (back to 100 thousand years). The record from any one such core does not usually depict global climate; however, suitable cores yield climate information that is representative of a wide region and may approximate a global pattern

Based upon these and other natural planetary archives, palaeoclimatologists and palaeoceanographers have established a sound understanding of the natural patterns and some of the mechanisms of past climate change (for example, Figs. 2, 5). This understanding provides the vital context against which contemporary changes in climate, including alleged human-caused warming, must be assessed.

What is a proxy record of temperature?
Tree rings and chemical indicators can represent past temperature.

The rate of tree growth and the ratios of different chemical constituents in ancient samples can vary with the prevailing temperature. Because of this, measurements made on modern materials, for example tree ring widths, can be used to establish the relationship between a set of particular measurements and the  range of conditions under which the selected objects grew or were deposited. Such relationships are called empirical, which means based on observation or experiment. Repeating the measurement on an ancient sample of the same material, and comparing the result with the established modern relationship, can then be used to infer the conditions that applied during the growth of the ancient sample.

Such measurements are called proxies, and a well known example is the relationship between the width of annual tree rings and the temperature for the year that each ring represents. Of course, tree ring width is also affected by other environmental variables such as annual rainfall and atmospheric carbon dioxide content. Such proxy indicators are not as sensitive to changing temperature as are modern thermometers, and do not provide a perfect reconstruction of past temperatures; nevertheless, they provide very useful information about past temperature and climate.

Perhaps the single most important proxy indicator of past temperature, not least because it can be determined for a wide range of ancient materials, is the ratio of heavy to light oxygen isotopes present.[10] Oxygen is present in ice (frozen H_2O), in ice bubbles (O_2 in trapped air) and in fossil shell material (calcite, or $CaCO_3$). The relevant isotope measurements, and hence inferred temperature variation, can therefore be derived from samples from both glacial ice cores and marine sediment cores. The well-established relationship that exists between the oxygen isotope ratio and the atmospheric or sea-water temperature at the time of deposition can then be used to determine the temperature at the time that the ancient sample was deposited.

Figs. 2 and 5 represent typical proxy climatic records that have been reconstructed using oxygen isotope measurements in this way.

How do we measure modern temperature?
In many ways: including with thermometers, ocean buoys, weather balloons and satellites.

The first primitive mercury or alcohol thermometers, called thermoscopes and lacking an accurate scale, were invented in Italy in the late 16th century. In 1612, the Italian inventor Santorio became the first person to add a scale to the measuring device, but it was not until one hundred years later, in 1714, that the German scientist Daniel Fahrenheit assembled the first mercury thermometer with the accurate scale that later came to be named after him.

Following these developments, systematic temperature observations by individuals and university staff were made from the middle of the 18th century onwards (see Fig. 10, p.76). But, of course, calculating a global average tempera-

---

10. Isotopes are forms of an element that are chemically identical but have differing atomic weights because of differing numbers of the subatomic particles called neutrons in their nucleus. The two commonest natural isotopes of oxygen are ^{18}O and ^{16}O. When physical processes occur, for example evaporation or rainfall, their differing atomic weights may result in the concentration of one or the other isotope, with the consequence that the ratio of the two isotopes changes from the average natural ratio. By measuring the changing isotope ratio, it is possible to infer how past temperature changed.

ture requires accurate observational records not just from a single place, but from a suitably located worldwide network of reliable thermometer observing stations. It was not until the middle of the 19th century that governments began to take on the role of organising and maintaining systematic meteorological observation networks. Meteorologists agree that an adequate network of such stations was not available before then, which is why the global average temperature records used in the global warming debate start at about that time.

An understanding of prevailing weather conditions is critical for the timely and safe passage of shipping, and routine meteorological observations have therefore long been a standard entry in ship's logs. In addition to air temperature on the deck, the ocean temperature was measured, initially on water collected in buckets from slow moving sailing ships and later from engine intake water on steam ships.

Thermometer measurements, then, are the first way in which quantitative records of temperature became available, and we now have about a 150 year-long historical record of determinations of average global temperature (Fig. 1, p.17).

Weather station thermometer records only provide us with temperature observations at ground level. As aviation developed after the World War II, it was recognised that a knowledge of the changeable winds and temperatures of the atmosphere was important for the safety of flight operations. Accordingly, during the 1950s national meteorological services began to implement systematic measurements through the atmosphere using thermistor[11] sensors that form part of radiosondes[12] on weather balloons. Since 1958, these balloons have been released twice daily at a network of about 900 locations throughout the world to measure a vertical profile of temperature. Interestingly, when combined into a global average the weather balloon record exhibits little overall warming between 1958 and 2002, but rather comprises a period of gentle cooling (1958-1977) followed by a slightly greater warming (1977-1998) (Fig. 9, upper). It is noteworthy, however, that the warming after 1977 is not manifest as a trend in the temperature record, but rather as stepped increases that occurred in 1977 and across the turn of the 21st century. Thus no significant linear increase in temperature occurred during a 54 year-long period during which atmospheric carbon dioxide increased from 315 to 394 ppm (25%).

[11.] Thermistor is a shorthand term for thermal resistor, a ceramic or polymer electrical component whose resistance varies in relation to temperature, making it suitable for accurate temperature measurement.

[12.] A radiosonde is the small instrument package suspended from the weather balloons that measure a vertical profile of temperature, relative humidity, pressure and wind direction and speed, and radio the data back to a base station.

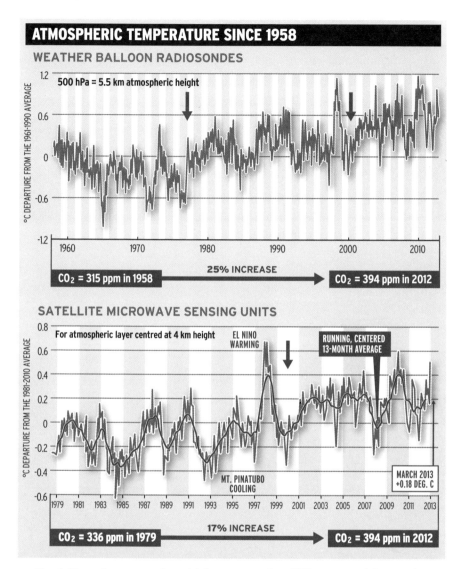

Fig. 9. Upper: Lower atmosphere global temperature since 1958, as measured from weather balloons (HadAT2, British Meteorological Office; Thorne et al., 2002). Note the presence of cooling from 1958 to 1977, followed by a warming of slightly greater magnitude between 1977 and 2012.

Lower: Lower atmosphere global temperature since 1979, as measured by microwave sensing units from satellites (NOAA; Spencer, 2013). For the period of overlap, the two records match well. But note that the overall warming that occurs between 1958 and 2012, across the two records, is represented not by a linear trend but by two step increases in temperature, in 1977 (Great Pacific Climate Shift) and around 2000 (downward-pointing red arrows).

Since 1979, a third method of measuring temperatures through the atmosphere has been developed using microwave sensing units (MSU) mounted on orbiting satellites. The MSU sensors accurately record the average temperature of a layer of the atmosphere by measuring the brightness of emissions from atmospheric oxygen molecules, which varies directly with changing temperature in a known way. The global average temperature record of the lower atmosphere calculated from the MSU data (Fig. 9, lower, p.75) corresponds well with the parallel ground thermometer and weather-balloon thermistor results.

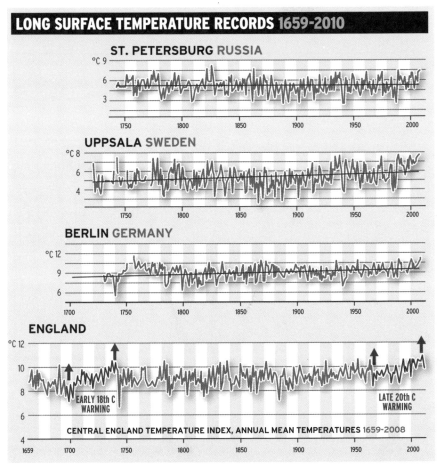

Fig. 10. Long surface temperature thermometer measurements starting in the early 18th century from selected Northern Hemisphere locations (above), compared with the longest record of all — the Central England Temperature index for 1659-2008 (below) (Willmore, 2009). All stations record a gentle warming trend over the last 300 more years, consistent with recovery from the Little Ice Age conditions, and show no particular influence from post-1950 carbon dioxide emissions. Note that the late 20th century warming in England had the same longevity, and proceeded at about the same rate, as the natural warming in the early 18th century (periods delimited by red arrows).

For the common period 1979-2011, all three temperature records show similar and significant year-to-year variability, especially for El Nino and La Nina events. For example, the rise and fall that preceded and followed the strong 1998 El Nino event was about 0.9°C. The magnitude of such year-to-year variability is large compared to the recent warming trends that are claimed as evidence for AGW (about 0.16°C/decade and 0.14°C/decade for thermometers and MSU respectively), which reduces the confidence that can be placed in the magnitudes of the trends. Furthermore, all such trends lie well beneath the warming of 0.2-0.3°/decade that the IPCC asserts should be caused by increases in atmospheric carbon dioxide.

How long is the record of direct measurements of temperature?

Up to 353 years at individual stations; about 150 years for a global network of stations.

We learned in the last answer that a satisfactory global network of weather stations didn't emerge until about 1860. This defines the last 150 years as the period of time over which we can calculate a useful curve of global average temperature using thermometer data (Fig. 1, p.17). However, earlier measurements are available for some individual places in the northern hemisphere, and can be used to reconstruct somewhat longer temperature histories (Fig. 10, p.76).

The longest established ground temperature record, termed the Central England Temperature Index (CETI), starts in 1659, which was soon after the invention of the thermoscope but before the Fahrenheit scale came into use. This 353 year-long data set (Fig. 10, lower) is archived by the British Meteorological Office, and was analysed recently by Scottish scientist Wilson Flood. Because of the depressed temperatures that occurred in the late 17th century as part of the celebrated Maunder Minimum[13], the overall CETI record does demonstrate a slight overall warming since then. Slight long-term warming trends are also present in seven other long northern hemisphere records that date back to the late 18th or early 19th century, three of which are also illustrated in Fig. 10. It

---

[13]. The Maunder Minimum, 1645-1715, was a period of intense cold that marked one of the minima of the Little Ice Age, and which occurred in association with a lack of sunspots and other solar activity at the time.

should be noted, however, that none of these records has been corrected for the Urban Heat Island effect.

When the CETI record is examined more closely, however, Wilson notes that:

> *The average CET summer temperature in the eighteenth century was 15.46°C while that for the twentieth century was 15.35°C. Far from being warmer due to assumed global warming, comparison of actual temperature data shows that UK summers in the twentieth century were cooler than those of two centuries previously.*

It seems that our longest available thermometer records, like our shorter and more accurate modern measures of temperature, offer little by way of evidence for the occurrence of dangerous human-caused global warming.

Over what time periods does temperature change reflect climate change?

30 years represents a Climate Normal, or one climate data point.

We learned earlier that climate is defined as being the average of a 30 year-long period of weather record (I: What is the Climate Normal?). Therefore, measurements over a 30 year period are needed to establish one climate data point.

In terms of measured temperatures, we have available to us some records from individual ground stations up to 353 years-long, and global average temperature records of length 150 years (ground thermometers), 54 years (weather balloon thermistors) and 33 years (satellite MSU). In terms of climate records, these datasets equate to 12, 5, almost 2 and 1 climate data points, respectively.

Therefore, important though they are in their own right, none of these in-

strumental records can adequately serve as a context for the study of climate change. Proper context is only provided when the short instrumental records are studied in the wider perspective of geological proxy records of climate change, which comprise at least hundreds of data points and stretch over thousands to millions of years.

We have already seen that such records for the last 6 million years reveal the presence of a steady background beat of longer-term climatic cyclicity (Fig. 2, p.17). For example, the long period of cooling that commenced about 3.5 million years ago was accompanied throughout by background climatic oscillations of 20,000, 41,000 and, more recently, 100,000 years in length (see below: What are Milankovitch variations?). And, on shorter, more human, time scales the significance of the 20th century warming is well shown in natural context by the Holocene part of the Greenland Ice Core record (Fig. 5, p.29).

What are Milankovitch variations?
Changes in Earth's orbit that control three fundamental frequencies of climatic oscillation.

The three fundamental frequencies of longer term climatic oscillation of 20,000, 41,000 and 100,000 years in length are termed Milankovitch frequencies. They are named after German meteorologist and geophysicist Milutin Milankovitch who, early in the 20th century, spent almost 20 years laboriously calculating the first graphs of Earth's recent orbital,

and coincidentally, climatic, history. Milankovitch's key insight was to understand that the distribution of solar radiant energy received across planet Earth changes through time in correspondence with geometric fluctuations that occur in the Earth's orbit, thus controlling seasonality and apparently the growth and decay of ice sheets. Though this insight has survived the test of time, it is understood today that it is the rate of change of the Milankovitch parameters, rather than their exact magnitude at any one time, that is most tightly coupled with climatic variation.

An interesting fact regarding the last few million years of climate history (Fig. 2, p.17) is that after about 3 million years ago the amplitude of the major oscillations increased at the same time as successive glacials got colder, whereas interglacial peaks tended to not exceed an upper boundary value that was only a little cooler than the preceding warm period in the Pliocene. At Earth system level, this is consistent with the existence of an assembly of negative feedbacks that together provide a warm-limit thermostat for global temperature.

The Milankovitch orbital variations are caused by gravitational interactions between the Earth and the other planets of the solar system, and affect both the tilt of the Earth's axis and the shape of its orbit around the Sun (Fig. 11,

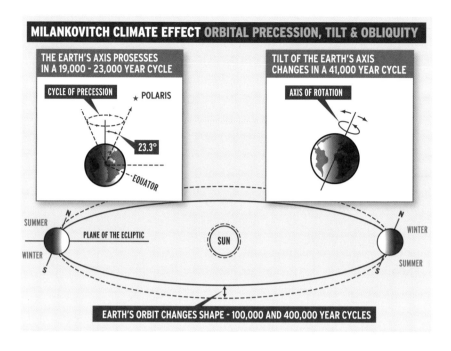

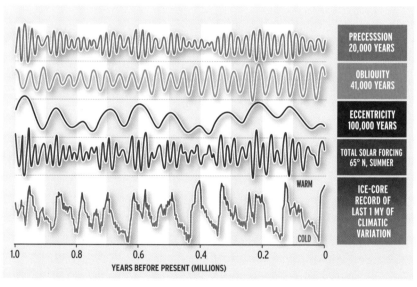

Fig. 11. Upper: Schematic of the changes in Earth's orbital characteristics around the sun that exert influence over climate, namely the 20,000 year (precession), 41,000 year (tilt) and 100,000 year (eccentricity) Milankovitch cyclicities (after a figure in New Scientist, 1989).

Lower: Pattern of Earth orbital variations and solar radiation forcing over the last million years. Precession (green), obliquity = tilt (orange) and eccentricity (navy blue) are combined in the red curve, which represents the summer forcing at latitude 65° N that is thought to play a large role in northern hemisphere glacial cycling. For comparison, the bottom curve (graded between red and blue grey) depicts the climate record represented by an Antarctic ice core.

upper). Specifically, the path of the orbit varies from more to less elliptical on a 100,000 year scale; the tilt of the Earth's axis varies slightly, between about 22.1° and 24.5° on a 41,000 year cycle; and finally the Earth's tilted axis also precesses ('wobbles' like a spinning top) on a roughly 20,000 year cycle.

Milankovitch's calculations enabled scientists to produce spectacular graphs such as Fig. 11 (lower), in which the various orbital characteristics can be projected backward in time. A full record of this predicted climatic cyclicity was first captured in the 1970s in deep ocean sediment cores (Fig. 2, p.17); and shortly thereafter confirmed in spectacularly parallel climate records from glacial ice cores in Greenland and Antarctica (compare Fig. 10, p.76, lowest plotted record).

These facts demonstrate the need to assess modern climate change against the pervasive natural climate cyclicity that already exists. On the longer time scale, the Milankovitch periodicities are the important ones, but other natural cycles shorter than 20,000 years also control climate variability. Most of the shorter cycles appear to be solar in origin, and many carry names. Important cycles include those with periodicities of 1,000-1,500 (Bond Cycle), 400, 180-210 (deVries or Suess Cycle), 70-100 (Gleissberg Cycle), 22 (Hale Cycle) and 11 (Schwabe Cycle) years, the latter representing the well known Sunspot cycle.

Does melting ice mean that global warming must be occurring?
Not necessarily.

Ice is naturally present on the Earth's surface in three main geomorphic forms: land-based ice-caps (Greenland and Antarctica), mountain-valley glaciers, and floating sea ice. Today, about 30 million km³ (91.7%) of all land ice is in Antarctica, 2.6 million km³ (7.9%) in Greenland and all other valley glaciers added together have a volume a little less than 0.1 million km³ (0.3%).

The implications of these figures are that, despite rousing media coverage, melting valley glaciers contain just 25 cm of global sea-level change equivalent, and are therefore largely irrelevant to concerns about sea-level rise. Major sea-level change from melting ice depends much more on the relative ice balance of the large Antarctic (82 metre sea-level equivalent) and Greenland (7 metre sea-level equivalent) ice caps, both of which appear to be in roughly steady state balance at the moment.

Icecaps and mountain glaciers

Icecaps and valley glaciers melt all the time around their edges or at their terminus, respectively. At the same time, precipitation of new snow proceeds over the inner or upper parts of the ice mass, and turns into the layers of new ice that sustain the outward flow of an ice-cap and the downslope flow of a valley glacier. When a balance exists between the amount of new ice accreted and the amount of old ice melted around the periphery, then the edge of an ice-cap or the terminus of the glacier will remain static in one place. Observations show, however, that such balance is an unusual happenstance. Over periods of decades to millennia, most ice masses either expand in size outwards or downslope (meaning that internal accretion must be exceeding peripheral melting) or alternatively shrink in size or retreat up valley (meaning that melting must be exceeding accretion).

Few observations of glacier extent exist prior to about 1860, though some inferences about earlier advances and retreats can be made from paintings, sketches and historical documents. Since 1860, however, many glaciers in the European Alps have been in a phase of steady retreat. Now, 150 years later, that retreat has revealed sub-fossil wood and *in situ* tree stumps, and also human artefacts and dwellings, which indicate that in earlier historic times the glaciers were smaller and situated further up their valleys, with normal vegetation and habitation immediately down valley.

Was this retreat driven by rising temperature? To some degree, perhaps, because the global temperature has certainly increased since the end of the Little Ice Age in 1860. However, reduced precipitation in the valley heads provides another equally plausible explanation for the glacial retreat, and it is most probable that both reduced precipitation and an increasing temperature have played a part in many glacier retreats.

Was this retreat driven by human-caused global warming? For the most part certainly not, though it is possible that human activities may have had a small influence over the last few decades. The reason is that the glacial retreat had been underway for fully 100 years (until about 1960) before the amount of human-related carbon dioxide emissions reached a level where they could conceivably have begun to raise global temperature enough to assist melting.

Sea ice

Sea ice expansion is driven by the spontaneous freezing of sea water in winter in areas of open polar ocean. Then, during the spring and summer months and as daily solar radiation increases with higher Sun angle and longer day length, the sea ice melts and the area contracts. This annual cycle results in changes in area of sea ice of about 10 million km^2 each year in the Arctic and about 12 million km² in the Antarctic (Fig. 12).

The formation of sea ice follows the annual cycle of heat loss by rad during the winter months followed by excess solar insolation in summer. The annual areal extent of sea ice is certainly influenced by temperature, both of the ocean and the atmosphere, and in general the colder that any winter is the more sea ice that will form. The melting and break-up of sea ice is, however, more complex, in that winds and ocean currents often play a major role in breaking up, dispersing and diminishing the area of sea ice, as was the case in the extensive diminution of ice that occurred in the Arctic Ocean in both 2007 and 2012.

We see that, as for land ice, the formation and melting of sea ice is not just a simple function of temperature, but reflects complex changes in a number of environmental variables. The satellite observational record of sea ice spans only 1979-2011, and shows an increase in area around Antarctica and a general decrease in area in the Arctic Ocean — which adds up to little net change in the overall global area of ice over the last 22 years. But 22 years does not even amount to one climate data point (I: What is the Climate Normal?), and is

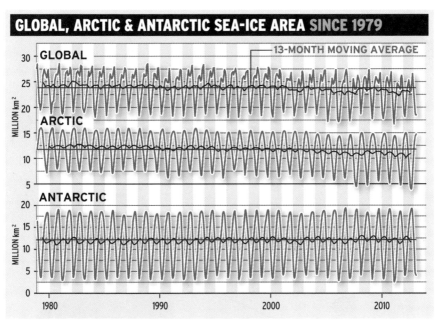

Fig. 12. January global sea ice area, 1979-2013 from satellite observations (Humlum, 2013, after NSIDC). Global ice area (top), northern hemisphere ice area (middle) and southern hemisphere ice area (bottom). In recent years a lower than long-term average area of sea ice in the Arctic (middle) has been mostly compensated for by a slightly higher than usual area of Antarctic sea ice (bottom). As a result, global sea ice cover is currently running close to the long-term average (top).

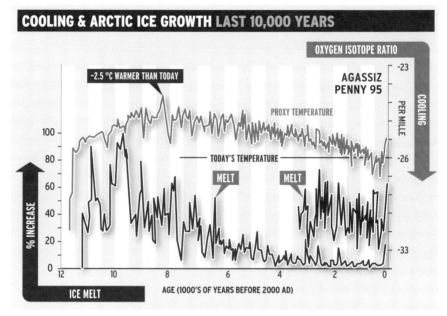

Fig. 13. Ice-melt in Arctic Canada for the last 12,000 years, reconstructed from summer melt-layers in two ice core records (Agassiz and Penny) (Fisher et al., 2006). Note the almost complete melting in summers during the early Holocene climatic optimum (9,000-10,000 years ago), and the steady decrease in melt since then, consistent with the long term cooling trend delineated by the oxygen isotope record (upper curve). These temperature changes, and meltings of ice-cap layers, would have been accompanied by parallel changes in nearby sea-ice cover.

therefore far too short a period of record from which to draw any meaningful conclusions about climate change.

Longer historical records demonstrate that the area of Arctic sea ice has fluctuated in a multi-decadal way in broad sympathy with past cycles in temperature, including shrinking to an area similar to that of recent years during periods of relative warmth in the 1780s and 1940s. Geological records show that earlier still, about 8,000 years ago during the early Holocene Climatic Optimum, temperatures up to 2.5° C warmer than today resulted in an almost or completely ice-free Arctic Ocean (Fig. 13).

Despite the enormous amount of media coverage over the last few years about the loss, or impending loss, of sea ice from the Arctic Ocean, no evidence exists of any recent changes that lie outside the range of natural climate cycling.

Finally, when both Arctic and Antarctic sea ice are considered together to produce a global estimate of sea ice cover, it is apparent that despite strong short-term fluctuations, little overall change has occurred in the long-term mean sea ice area for the last 42 years (Fig. 12, p.83).

What about other circumstantial evidence; coral bleaching or polar bears anyone?

Corals bleach and polar bears vary in number, but neither is evidence for man-made global warming.

The advance or retreat of glaciers, is but one of the many lines of 'evidence' advanced in favour of the occurrence of global warming. Other changes in the natural world commonly attributed to human-caused warming include such things as coral bleaching episodes; fewer polar bears; birds nesting earlier, birds nesting later; more droughts, fewer droughts; more floods, fewer floods; more hurricanes, fewer hurricanes; and so on.

Even when such changes are accompanied by a rising temperature (which is often claimed but not always the actual case), the mere existence of such parallel changes says nothing directly about their cause, which means that such events cannot provide direct evidence for human causation. This is, first, because all changes that have been reported fall within the boundaries of previous natural variability; and, second, because other changes that are not publicised are currently proceeding in the opposite direction to that expected in a warming world.

In particular, and despite the widespread alarmism raised in the media, the area of sea ice in the Arctic Ocean is not unusually small and the global sea ice cover is not decreasing rapidly (Fig. 12, p.83), the number or intensity of tropical storms is not increasing (Fig. 14, p.86), the rate of sea-level rise is not accelerating (compare Fig. 25, p.135) and the number of polar bears is not decreasing. An excellent and independent analysis of these and other climate-related scares, with many references, is provided by the 2009 Report of the Non-Governmental Panel on Climate Change (NIPCC).[14]

In Australia, coral bleaching on the Great Barrier Reef and phases of drought, bushfire and flood (especially in the iconic Murray-Darling river system) are commonly asserted to be linked to human-caused global warming. No substantive research results support any of these claims. Instead, many recent research articles have demonstrated that Australia's year-to-year weather variability, and the occurrence of extreme

[14.] *http://www.nipccreport.org/reports/reports.html*

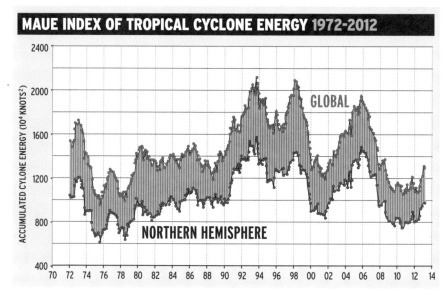

Fig. 14. Summary plot (2-year running sums) of the total energy contained in tropical storm systems, 1972-2013 (Maue, 2013); upper curve, global; lower curve, Northern Hemisphere. No long term trend exists, and tropical storm activity has been at a very low level over the last few years. The peak in activity in 2005 included US hurricane Katrina.

events such as floods and droughts, mostly occur in response to changes in two influential climatic oscillations, namely the El Nino-La Nina-Southern Oscillation cycle (ENSO, originating in the Pacific Ocean) and the Indian Ocean dipole (IOD, originating in the Indian Ocean) (see VIII). These recurring changes in ocean surface temperature patterns are an outcome of varying ocean current circulations.

All the matters mentioned above are, of course, proper topics for investigative research. But despite the recurring alarm generated by media coverage, no study to date has established a certain link between changes in any of these things and human emissions of carbon dioxide

It repeatedly escapes influential public commentators, such as Mr Al Gore and the government's Climate Commissioners, that the Earth is a dynamic planet. Earth's systems are constantly changing, and its lithosphere, biosphere, atmosphere and oceans incorporate many complex feedback buffers. Changes occur in all aspects of local climate, all the time and all over the world. Geological records show that climate also changes continually through deep time.

Change is what climate does, and the ecologies of the natural world change naturally and concomitantly in response.

What is the Holocene and why is it important?
The period of climatic warmth (interglacial) that elapsed over the last 11,700 years.

The peak of the last great Pleistocene ice age occurred about 20,000 years ago (Fig. 2, p.17). Over the ensuing 8,000 years climatic warming caused the ice-sheets to melt, glaciers to retreat and sea-levels to rise (Fig. 6, p.31).

The Holocene is the name given to the post-glacial warm climatic interval in which we now live, which commenced about 11,700 years ago during the cultural period called the Mesolithic. During the Mesolithic and following Neolithic periods, *Homo sapiens* discovered how to make pottery, domesticate animals and cultivate crops, how to smelt first bronze and then iron, and how to develop city civilisations — many of these developments surely being aided by the relative warmth and accompanying wetness of the climate at the time.

The scientific importance of the Holocene is that it is the climatic state that accompanied the emergence of human civilisations, the global temperature pertaining not being very different from the other cyclic warm peaks of the last 3 million years. The climatic record of the Holocene is therefore the benchmark against which recent climate variations need to be compared.

Excellent Holocene climatic records are available from ice cores, one from Greenland having been discussed earlier (Figs. 5, 6). This record demonstrates three critical facts. First, that the long term temperature record of the Holocene in the northern hemisphere has been one of cooling by about 2°C from a climatic optimum in the early Holocene. Second, that throughout the Holocene a 1,000-1,500 year-long climate cycle of 1-2.5°C magnitude (the Bond Cycle) has been conspicuous, and that we live today in the latest warm peak of that cycle. And third, in comparison with the full Holocene record the temperatures of the late 20th century were not unusually warm.

In Greenland, the three most recent historic warm peaks of the Bond cycle (Minoan, Roman and Medieval Warm Periods) all attained or exceeded the magnitude of the late twentieth century warming. Many other records from around the world confirm the greater warmth of the Medieval over the Late 20th Century Warm Period — a critically important fact because it underscores that there is nothing unusual about

MT. EVEREST 7000 B.C.

the warming experienced towards the end of the 20th century.

Late 20th century warmth simply reflects the most recent peak of the 1,000 year climatic rhythm, which in itself is a modulation of the declining average temperature of the Holocene interglacial in similar fashion to that which occurred in earlier interglacials.

What were the Medieval Warm Period and Little Ice Age?
Significant warm and cold climatic intervals that occurred over the last 1,000 years.

In the absence of instrument measurements, scientists rely upon historical and geological outcrop data to provide estimates for temperatures over historic times.

Such information, including for example the known cultivation of vineyard grapes in the north of England around 1,000 AD and the widespread erection of monumental buildings through Europe over the following few centuries, suggests a period of warmth and prosperity. Conditions deteriorated subsequently, and intensely cold winters with many reports of frozen rivers characterised European climate for the next several hundred years. The marked climatic deterioration from the end of the first millennium led to the naming of a major warm episode around 1,000 AD as the Medieval Warm Period (MWP), and of the prolonged, intermittent cold spell between about 1350 and 1860 as the Little Ice Age (LIA) (Fig. 5, p.29).

The flawed statistical analysis of proxy records that led to the 'hockey stick' representation of the last 1,000 years of temperature (II: What is the hockey-stick and why was it important? Fig. 8, p.51) led IPCC scientists to allege in 2001 that the MWP and the LIA climatic intervals were confined to Western Europe, and regional only. In reality, proxy data from many parts of the world give strong credence to the Medieval Warm Period and Little Ice Age being global, although naturally their characteristics vary regionally in nature and strength. For example, the carbon dioxide record recovered from Law Dome, Antarctica shows a minimum during the 17th century, consistent with the depth of the Little Ice Age as documented over Europe.

Today, many research papers continue to document the existence of both the Medieval Warm Period and the Little Ice Age in widely distributed, far field locations from around the world, as is well documented by listings at the CO2 Science website. More and more proxy data such as oxygen isotope records are filling the gaps and supporting the conclusions drawn previously from historical records and geological interpretations.

IV

THE GREENHOUSE HYPOTHESIS

Is the Earth in climatic equilibrium?
Yes and no.

The Sun is the primary source of energy that heats the Earth. Incoming solar radiation carries most of its energy as shortwave radiation, which comprises both invisible ultraviolet (0.01-0.4 μ[15] wavelength) and visible light (0.4-0.7 μ wavelength). The amount of energy received at the top of the atmosphere varies between about 1340 watts/m^2 at the equator and zero at the poles, for a global average of 340 watts/m^2. These numbers can be understood in the context that a small domestic bar heater emits a total of about 1000 watts.

In order for energy balance of the Earth to be achieved (i.e. to avoid either heating or cooling over time), the average 340 watts/m^2 of incoming energy must be offset by the emission of a similar amount of radiation energy back to space. This emission occurs as longwave, or 'Earth', radiation in the infrared (0.7-1000 μ wavelength) portion of the radiation spectrum.

[15.] 1 micron (μ) = one one-thousandth of a mm, or 0.001 mm

Earth, however, is never in exact radiation balance, because the global temperature changes with season and from year to year. Nonetheless, the relative stability of global temperature that is observed over decades and centuries means that over these timescales near radiation equilibrium is achieved.

The transmission of energy back to space takes place by two main processes. The first is the direct reflection of incoming sunlight by light-coloured materials such as clouds or glaciers. The lighter-coloured the material (scientists say, the higher its albedo), the more energy that is reflected; dry, powdery snow, for example, can reflect more than 80% of incoming sunlight. The second process of heat loss is by emission of infrared radiation, as we have already described, i.e. by the same mechanism by which our imagined bar heater operates.

Not all of the incoming solar radiation is absorbed at the Earth's surface, and nor does all of the radiation to space emanate from the Earth's surface itself. Despite their high reflectance, some of the solar radiation is absorbed by clouds, and in clear skies another fraction is absorbed by the gases and aerosols of the atmosphere. Nonetheless, the major fraction of solar radiation not reflected directly to space is absorbed by the land and water that makes up Earth's surface.

The radiation emitted back to space emanates, then, from the surface, from clouds, and from various greenhouse gases in the atmosphere. However, only a relatively small fraction of the radiation emitted from the Earth's surface is able to reach space directly, because most is absorbed by the overlying clouds and greenhouse gases of the atmosphere.

This greenhouse effect, which is explained in more detail next, forms a key role in keeping the Earth's surface warmer than is required solely for balancing surface emission with the Sun's incoming radiation.

What is a greenhouse gas?
A gas that both absorbs and emits longwave radiation.

Earth's well-mixed, dry, gaseous atmosphere comprises 78% nitrogen (N_2), 21% oxygen (O_2) and a combined 1% of trace gases, of which the greater part (0.9%) comprises argon (A) and carbon dioxide (at .039%, or 390 ppm).

As Earth's outgoing radiated infrared energy, carried by electromagnetic photons[16], traverses the atmosphere, some of the photons collide with molecules in the air. For the dominant constituents of the atmosphere, nitrogen and oxygen, such collisions simply result in deflections of the photon and

---

[16] A photon is an elementary particle, and comprises the quantum of light and all other forms of electromagnetic radiation.

the molecule. But for a minority of atmospheric gases, an incident photon, if of the appropriate wavelength, is absorbed by the molecule which thereby has its energy increased. Such gases are called greenhouse gases, and the most important, in order of magnitude of their overall contribution to greenhouse warming, are water vapour (H_2O), carbon dioxide (CO_2) and methane (CH_4).

Greenhouse gases emit infrared radiation in all directions at the same characteristic wavelengths that they absorb. According to well established laws of physics the intensity of emission is a function of the absolute temperature[17], and warmer gases emit more intensely than cooler gases. The net radiation transfer up through the atmosphere of the finally outgoing energy corresponds to the summation of all the absorptions and emissions by the individual greenhouse gases along the way.

The radiation emitted back to space is of longer wavelength than incoming radiation because of the characteristic temperatures of the components of the climate system. For example, temperatures of the high atmosphere and at the polar surface in winter are typically colder than -20°C, whereas tropical ocean surfaces are typically about 30°C and tropical land surfaces may reach more than +50°C. Each of these different regions emits radiation of the magnitude and wavelength that is characteristic of its temperature.

When viewed from space, the radiation leaving Earth emanates from various altitudes according to the distribution of each greenhouse gas and its characteristic wavelength. For those wavelengths associated with carbon dioxide and ozone, the emission is from the high atmosphere (the stratosphere), whereas those wavelengths associated with water vapour are emitted from the low and middle atmosphere (the troposphere). At some wavelengths, known collectively as the 'atmospheric window', no common greenhouse gas absorbers/emitters exist and radiation is accordingly emitted directly to space from Earth's surface and the highest cloud tops.

A greenhouse gas, then, is simply a gas that has the property of absorbing and emitting infra-red radiation. Such gases are important for Earth's climate because by regulating the transfer of energy through the atmosphere they play a key role in controlling temperature.

[17.] The Absolute Temperature scale is a measure of internal energy and is given in degrees Kelvin where 0°K is equivalent to -273°C and the scale units are equivalent to degrees Celsius.

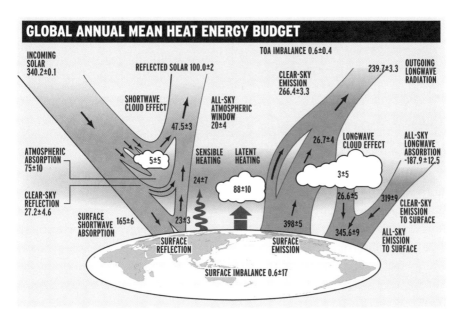

Fig. 15. Global annual mean energy budget of Earth for the period 2000–2010 (Stephens et al., 2012). All fluxes are in watts/m². Shortwave solar radiation fluxes are in blue-gray and longer wave infrared fluxes in rose pink. See text for more detailed explanation.

Because the emission of energy by a molecule of greenhouse gas occurs in all directions, half of the total is directed downwards towards the Earth's surface, and it is this energy that contributes to the heating of the lower atmosphere called the greenhouse effect (below: What is the classical greenhouse effect). The portion of the emitted energy that is directed upwards and eventually escapes to space is what balances the absorbed solar radiation.

The Sun obviously warms the Earth, but what cools it?
Amongst other things, greenhouse gases.

As explained before (Is the Earth in climatic equilibrium?), the energy stability of the Earth is maintained by balancing the incoming solar energy by the reflection or retransmission of an equivalent amount of energy back to space.

A recent summary paper by Graeme Stephens and co-authors estimates that the fate of the incoming average 340 watts/m² (100%) of solar energy at the top of the atmosphere is as follows (Fig. 15, left hand side, blue-gray colour hue):

75 watts/m²   (22%)   absorbed in the atmosphere.
165 watts/m²   (49%)   absorbed by the Earth's surface.
100 watts/m²   (29%)   reflected back to space by atmosphere & surface.

The atmosphere and Earth (the climate system), having gained heat through the absorption of 240 watts/m² (71%) of incoming solar radiation, now seek to restore the overall energy balance by emitting heat back to space. Complex energy exchanges occur that involve back-radiation loops from the atmosphere, including importantly clouds and greenhouse gases, but the ultimate result is that 240 watts/m² of heat energy escapes to space as longwave radiation from the top of the atmosphere (Fig. 15, right hand side, rose pink colour hue). When added to the 100 watts/m² of incoming solar energy that is directly reflected, this escaping radiative energy balances the incoming energy (240+100 = 340 watts/m²), and the overall planetary energy balance is sustained.

Thus the answer to the question posed at the beginning is that the energy that is directed back to space by direct reflection (29%) and by radiation (49%+22% = 71%) is what cools the Earth; and of the radiative loss more than half is energy emitted by atmospheric greenhouse gases, the remainder issuing from Earth's surface.

In other words, and to a degree counter-intuitively, greenhouse gases help to both cool (by the radiation that they emit to space from the high atmosphere) and warm (by the radiation that they emit down towards the lower atmosphere and surface) the Earth.

Is carbon dioxide the most important greenhouse gas?
No, water vapour is the most important greenhouse gas.

There are many naturally occurring greenhouse gases in the atmosphere, including water vapour, carbon dioxide, methane, ozone, and oxides of nitrogen. The presence of each is associated with physical or biological processes and their respective concentrations represent the net outcome of those processes.

Earth is the water planet. 70% of the Earth's surface is covered by ocean and evaporation is continually supplying water vapour to the atmosphere. Much of the rain falling on land is returned to the atmosphere either directly as evaporation or indirectly through plants by evapotranspiration. On average, water vapour spends less than two weeks in the atmosphere before it is drawn into clouds, condenses and falls to the surface as rain.

The rate of evaporation and the ability of the atmosphere to hold water vary directly with Earth's temperature. As Earth warms the atmosphere is able to hold more water vapour. However, as Earth warms the rate of evaporation and evapotranspiration also increase, which in turn increases the rate of removal of energy from the surface in the form of latent heat (see footnote 1, p. 20). Thus both 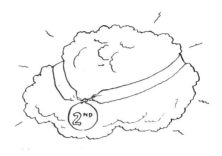 warming (extra greenhouse gas into the atmosphere) and cooling (evaporation) mechanisms are operative. Also, more water vapour in the atmosphere helps to create low cloud, which reflects incoming sunlight, an additional cooling effect. Earth's temperature remains relatively stable overall precisely because these various feedback processes are in near balance.

Carbon dioxide enters the atmosphere by exchange with the oceans and through decay of biological material. It is removed by exchange with the oceans and by growth of vegetation. Plants extract carbon dioxide from the atmosphere during photosynthesis. Most of the carbon taken from the atmosphere during photosynthesis is later returned, either as carbon dioxide or methane, as plant material decays. A small fraction is bound in the soil or ocean sediment and becomes fossilised as limestone or potential coal, oil and natural gas supplies.

Carbon dioxide has a residence time in the atmosphere of about 7 years, much longer than water vapour. The IPCC asserts a longer lifetime still of up to 200 years. However, this represents not the average residence time, but rather the time taken from injection of an additional amount of carbon dioxide into the atmosphere to when it returns to the original concentration. This IPCC 'lifetime' is used in calculating what is called the 'Global Warming Potential' of various greenhouse gases.

The natural sources and sinks of carbon dioxide are globally dispersed, and the gas is well mixed by atmospheric motions. As a consequence, its concentration varies little in space. The natural processes of the carbon cycle (the ocean-atmosphere physicochemical exchange processes, and the growth and decay of vegetation) are in near balance. So although it fluctuates with the seasons, in the short term the average atmospheric concentration of carbon dioxide does not vary much through time. Nonetheless, the natural processes do vary with Earth's temperature, and ice core analyses and other evidence shows that global carbon dioxide concentrations are higher when Earth is warmer and lower when Earth is colder.

The reason that carbon dioxide is the focus of so much attention is not that it is the prime greenhouse gas, for it isn't. Rather, the perceived problem is that modern industrial processes are consuming fossil fuels (coal, oil and gas) that have accumulated over millions of years and releasing the carbon dioxide back into the atmosphere at a fast rate. Over the past century an increase in atmospheric carbon dioxide from about 300ppm to near 400ppm has been attributed to the burning of fossil fuels, with nearly 50% of this increase occurring over the last two decades.

It is this steady increase in atmospheric carbon dioxide that is the basis for concern about human enhancement of the greenhouse effect. However, the magnitude of the warming that will be produced by the extra carbon dioxide remains much disputed amongst expert scientists. The empirical evidence indicates that any increased warming due to human-related carbon dioxide is very small and lies submerged within the natural variation of the climate system (I: Is dangerous warming being caused by carbon dioxide emissions?)

What is the classical explanation for the greenhouse effect?
That greenhouse gases absorb space-bound radiation from Earth's surface, causing the lower atmosphere to be warmer than we might otherwise expect.

The greenhouse effect is an important characteristic of the Earth's climate, but its cause and description are widely misunderstood.

As can be seen from Figure 15, the average radiation emission from the Earth's surface (398 W/m²) is much greater than the absorbed solar radiation (165 W/m²). An essential requirement for the Earth's average temperature to remain nearly constant, therefore, is that the ongoing rate of absorption of incoming solar radiation must be balanced by an equivalent rate of

loss of energy to space (above, Is the Earth in climatic equilibrium?). There is therefore a need to explain how the observed surface radiation imbalance is sustained, and why surface temperature is so much warmer than might be expected from the available solar heating.

This major dilemma was already apparent in the early 1800s, when the first rudimentary estimates were made of the surface temperature of the Earth and of the intensity of longwave radiation that this temperature should generate. The radiation intensity from the Sun was measured to be much less than the longwave emission to space should have been, given Earth's average surface temperature. Somehow energy appeared to be being retained by the Earth, to maintain a temperature higher than could be sustained by the magnitude of incoming solar radiation alone.

In the 1820s, the French mathematician Joseph Fourier provided the first plausible explanation for this energy budget dilemma that gained wide acceptance. Fourier's idea (hypothesis) was that the atmosphere contained gases that absorbed a significant part of the longwave radiation emitted by the surface, such that only a fraction of the surface energy ended up being directly emitted to space, the other fraction causing warming of the atmosphere. Calculating the energy absorbed in this way, and adding it to the direct radiation energy emitted at the top of the atmosphere, a balance can be achieved between the amount of solar radiation being received and the amount of longwave radiation being either absorbed in the atmosphere or emitted to space. The radiation absorbed in the lower parts of the atmosphere was suggested to cause heating there, thus resolving the initial observational conundrum.

The process whereby gases in the atmosphere maintain the temperature of Earth's surface and lower atmosphere as warmer than they would otherwise be has become known as the greenhouse effect. The analogy is with the warm, glass-enclosed plant conservatories called greenhouses, which combat the effects of frost on plant growth. The terminology is now entrenched, but nonetheless the analogy with a greenhouse is misleading — for a greenhouse warms mainly because its enclosure prevents the convective loss of heat that occurs in the free atmosphere.

Fourier's explanation gathered support from experimental work in the 1850s by the English physicist John Tyndall, who measured the attenuation of longwave radiation by the various gases present in air and found that water vapour and carbon dioxide were indeed powerful absorbers of radiation. Although only minor constituents of the atmosphere, it became clear that these two gases absorbed a significant fraction of Earth's outgoing longwave radiation. In contrast, the major constituents of air, oxygen and nitrogen, were found to make negligible contribution to the absorption of Earth's radiation emissions.

And thus it was that water vapour, carbon dioxide and other gases that absorb longwave radiation became known as greenhouse gases.

How is the greenhouse effect now understood by scientists?
As more complex than described by Fourier, because heat redistribution takes place by atmospheric convection as well as radiation.

Nearly 200 years later, Fourier's explanation of the greenhouse effect remains widely accepted, and as recently as 2007 the 4th Assessment Report of the IPCC explained the greenhouse effect in exactly such terms:

> 'Some of the infrared radiation passes through the atmosphere and some is absorbed and re-emitted in all directions by greenhouse gas molecules and clouds. The effect of this is to warm the Earth's surface and the lower atmosphere.'

Notwithstanding this endorsement by the IPCC, atmospheric scientists understand that a fundamental problem exists with the Fourier explanation, which, though correct so far as it goes, is an oversimplification of the real situation.

Atmospheric greenhouse gases don't just absorb longwave radiation but also emit it in all directions, some upwards towards space and some downwards towards Earth's surface. It has been recognised since at least the 1950s that the magnitude of longwave radiation emitted by greenhouse gases exceeds the radiation energy from the Earth's surface that the gases intercept and absorb. This means that there is an on-going net energy loss of longwave radiation by the greenhouse gases, which turns out to be equivalent to a rate of cooling of the atmosphere of more than 1°C per day.

Clearly, if the atmosphere is emitting more longwave radiation (in all directions, including both out to space and back to Earth) than it is absorbing from the Earth and directly from the Sun, then it cannot be greenhouse gases alone that are keeping the Earth's surface and lower atmosphere at a relatively elevated temperature.

The solution to this condundrum, and the kernel of a more accurate explanation for the greenhouse effect, was provided by US meteorologists Herbert Riehl and Joanne Malkus in 1958. These scientists noted that over the tropics the Earth's surface absorbs more solar radiation than the net energy that it loses by longwave radiation, whereas in contrast the tropical atmosphere emits more longwave radiation than it absorbs; the surface therefore gains radiation energy and tends to warm, while the atmosphere loses radiation energy and tends to cool.

The dilemma that then needs explanation is how energy is transferred from the surface to the atmosphere in order to offset the effects of its radiative cooling and to maintain an overall steady state. Two obvious candidate mechanisms for the energy transfer are conduction and turbulent heat transfer, but neither provides a satisfactory explanation. The first fails because air is a very good insulator against heat conduction. And the second fails because, although temperature decreases with increasing height in the atmosphere, the potential temperature (which represents the sum of internal energy and potential energy[18]) increases with height so that turbulence will actually transfer energy downwards in an opposite direction to that needed.

To solve the dilemma, Riehl and Malkus proposed that, over the tropics, excess surface energy is transferred to the atmospheric layer near the surface as both heat (by conduction) and latent energy (by evaporation). Much of the tropical surface is made up of oceans, thus facilitating evaporation of moisture as a major component of energy exchange. As the trade winds move the boundary layer[19] air towards the inter-tropical convergence zone near the equator, the exchange of heat and evaporation of moisture causes the total energy of the layer (the sum of the internal energy, the potential energy and latent energy) to rise.

The combination of this heating of the boundary layer and the radiative cooling of the wider atmosphere causes the boundary layer to become unstable in regions near the equator. Eventually the atmosphere becomes sufficiently unstable for boundary layer air to rise buoyantly into the atmosphere, becoming visible as deep convection cumulus and cumulo-nimbus clouds. The latent energy released by the water vapour condensing in the clouds thereby becomes available to offset the loss of radiation energy from the atmosphere generally, thus maintaining an overall energy balance.

An essential feature of the Riehl-Malkus explanation of the energy balance is that buoyant convection transfers heat and latent energy through the atmosphere to offset radiation loss. Such deep convection cannot take place until the atmosphere becomes unstable, which is achieved when the rate of decrease of temperature with height (termed the lapse rate) exceeds a characteristic value that in the tropics is near 6.5°C/km near the surface and rises to about 10°C/km in the high atmosphere.

---

[18] Internal energy is the sum of all forms of microscopic energy of a system, as represented by the disordered, random motion of molecules. Potential (or stored) energy is the ability of a system to do work by virtue of its position (think gravity) or internal structure (think coiled spring or electric charge).

[19] The boundary layer, typically a few hundred to 2000 metres in height, is that well mixed lowest part of the atmosphere that has its behaviour (wind speed, turbulence, moisture, temperature, etc.) influenced by the proximity of the Earth's surface.

Thus the radiation–convection model established by Riehl/Malkus explains the vertical temperature profile of the tropics as being regulated by the thermodynamics of convection. Solar heating of the surface and longwave radiative cooling of the atmosphere combine to increase the temperature lapse rate until buoyant convection is established, thereby distributing heat and latent energy from the surface through the atmosphere at a rate necessary to offset its radiative cooling — and thus regulating temperatures at the surface and through the atmosphere.

The greenhouse gases and clouds emit radiation to space across differing wavelength bands from altitudes that reflect the characteristics and concentration of each. But on average, across all wavelengths, the radiation that Earth emits to space originates from an altitude of about five kilometres — which is called the characteristic emission height. It is the temperature of the greenhouse gases and clouds here (the characteristic temperature) that determines the magnitude of the emission to space. Adding extra greenhouse gas to the atmosphere causes an elevation of the characteristic emission height, which maintains both the characteristic temperature and the emission to space.

The temperature lapse rate through the atmosphere as regulated by convection ensures that as the characteristic emission height rises, so too does the surface temperature. For every 100 metre rise in the characteristic emission altitude the surface temperature will rise about 0.65°C.

The vertical transfer of energy through the atmosphere by convection in the Riehl/Malkus fashion necessitates that the surface temperature is warmer than the characteristic emission temperature of the middle atmosphere. It is this characteristic that represents the greenhouse effect on planet Earth.

What is the greenhouse effect as understood by the public?
That dangerous global warming is being caused by human carbon dioxide emissions.

In general communication, and in the media, the terms greenhouse, greenhouse effect and global warming have come to carry a particular vernacular meaning almost independently of the scientific background provided in the two previous entries.

When an opinion poll or a reporter solicits information on what members of the public think about the issue they ask questions such as, 'do you believe in global warming', 'do you believe in climate change', or 'do you believe in the greenhouse effect'.

Leaving aside the issue that science is never about belief, and thanks to the UN's unique definition of global warming, all such questions are actually coded, and are understood by the public to mean 'is dangerous global warming being caused by man-made emissions of carbon dioxide'. Needless to say, this is a completely different, albeit distantly related, question.

These and other ambiguities ('carbon' for 'carbon dioxide', for example) are widespread in Australian and New Zealand political and media usage. They lead to great confusion in the discussion of climate change and its causes, and also undermine the value of nearly all opinion poll results. For there is no way of knowing how many of the persons answering 'yes' to the question 'do you believe in climate change' are giving the correct scientific answer (the 'yes' that would be provided by all six authors of this book), and how many have read the hidden code in the question and answered 'yes' because they are worried about warming caused by human emissions and think that that is what the question is asking about.

Is less warming bang really generated by every extra carbon dioxide buck?

Yes, carbon dioxide is of limited potency and waning influence as its concentration increases.

Carbon dioxide is a potent greenhouse gas for intercepting radiation across specific portions of the infrared spectrum, notably at wavelengths around 14.8 μ and 9 μ. Initially, at low atmospheric concentrations, the gas therefore has a strong greenhouse effect as it intercepts outgoing radiation at these wavelengths. However, the narrowness of the spectral intervals across which carbon dioxide intercepts radiation results in a rapid saturation of its effect, such that every doubling in the concentration of carbon dioxide present enhances the greenhouse effect by a constant amount. This is reflected as the negative logarithmic relationship that actually exists between extra carbon dioxide and the warming that it causes.[20]

[20.] The relationship between increasing carbon dioxide and extra radiation forcing is summarised by the IPCC (3AR, 2001, p. 358) as $\Delta F = \alpha \ln(C/C_o)$, where ΔF is the extra radiation forcing (W/m²); α is a constant termed the climate sensitivity (assumed by the IPCC to be 6.3); C is the actual concentration (today, 390 ppm), and C_o is the pre-industrial concentration (280 ppm) of carbon dioxide.

The dramatic effect that a logarithmic scale has on the changing magnitude of incremental changes in two variables is illustrated by Figure 16. The diagram displays a projection, using the MODTRAN stan-

dard atmospheric model (University of Chicago), of the increasing radiation forcing that occurs when carbon dioxide in 20 ppm increments is injected into Earth's atmosphere.

Starting at the left hand side, it is apparent that the first 60 ppm of carbon dioxide injected produces a strong cumulative radiation forcing of 19.9 W/m² (15.3, 2.9 and 1.7 W/m²) in a pattern that from the outset displays a rapidly declining magnitude of extra radiation forcing for each successive 20 ppm increment. This pattern continues as one moves across the figure to the right, such that even at the relatively impoverished carbon dioxide level of 180 ppm that marked recent glacial episodes, more so at the 280 ppm level that marked the pre-industrial atmosphere and even more so at today's 390 ppm, further 20 ppm increases in carbon dioxide produce only a tiny (0.2 W/m², and successively lessening) amount of extra radiation forcing.

Given where Earth's atmosphere sits on the scale today (at 390 ppm), it is apparent that further increases in carbon dioxide will produce only very small increases in radiation forcing and thus global warming. Regarding the much-feared doubling of the pre-industrial level (i.e. an increase from 280 to 560 ppm), and noting the decreasing radiation forcing inherent in the logarithmic relationship (Fig. 16, p.103), at 390 ppm the Earth has already realised nearly 50% of the additional radiation forcing and anthropogenic warming that will be induced by a full doubling of carbon dioxide from pre-industrial levels.

Given the importance of the less-bang-for-every-additional-buck nature of the relationship between increasing carbon dioxide and extra warming through radiation forcing, it is extraordinary that an explanation of the matter is almost entirely absent from the public debate. Even if all of the up to 0.8°C warming that occurred in the 20th century were to be attributed to carbon dioxide (an unsubstantiated and highly contentious proposition in itself), the relationship implies that a full doubling of carbon dioxide over pre-industrial concentrations will produce under 1°C of future additional warming. Further, if the concentration were to double again to 1120 ppm, the additional global temperature increase would be less than another 2°C (below: What net warming will be produced by a doubling of carbon dioxide?).

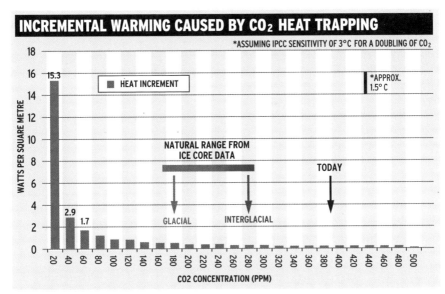

Fig. 16. Model projection of the incremental increases in radiative heat trapped in the lower atmosphere (rather than lost to space) by injections of carbon dioxide in 20 ppm increments (MODTRAN atmospheric model, University of Chicago). Calculations are in terms of watts/m^2 of radiant heat increase (left hand axis scale). Translating each increment of heat trapped in the atmosphere into degrees Celsius depends upon the assumed sensitivity of the climate system, which remains controversial (see Fig. 17, p.105). The approximate temperature bar (top right) is based upon the IPCC's estimated sensitivity of 3.3° C for a doubling of carbon dioxide. Note that this temperature increase, whatever its precise value, is a constant that applies to all doublings of carbon dioxide, for example from 140 to 280 ppm, 280 to 560 ppm and 560 to 1120 ppm.

What is climate sensitivity?

The degree to which temperature increases as carbon dioxide increases.

The climate sensitivity is defined as the amount of warming that will be produced by a doubling of carbon dioxide from its pre-industrial level of 280 ppm to 560 ppm.

Nearly all scientists agree that the effect of thus doubling carbon dioxide on its own will generate radiation forcing of about 3.6 W/m^2 that could produce a warming of up to 1°C. This is because, at the temperature of the Earth, laboratory experiments have established that a flat dry metal plate will emit additional longwave radiation of about 5.5 W/m^2 for every degree C temperature rise (the well-known Stefan–Boltzman Law of physics); prima facie, therefore, an additional 3.6 W/m^2 of heating might be expected to cause a temperature rise of about 0.7°C.

The Earth, however, is not a metal plate, and nor does it have a dry surface or dry atmosphere. That Earth is the water planet is an especial complication in relating laboratory physics to the real world. The evaporation of water and transfer of latent heat from the oceans, wet soils and vegetation all tend to lessen the amount of any surface temperature rise caused by increased carbon dioxide. However, other factors act in the opposite direction and amplify the initial warming effect of carbon dioxide. These lessening and amplifying factors are called negative and positive feedbacks, respectively.

For example, as the atmosphere warms it holds more water vapour, itself a powerful greenhouse gas. Then there are clouds that can either increase or decrease Earth's reflectivity (albedo), or warm or cool the surface, depending on their altitude and whether they contain water droplets or ice crystals.

As discussed next, many different opinions exist regarding the magnitude of the various feedbacks associated with doubled carbon dioxide, and whether they have a net positive or negative influence on temperature.

What net warming will be produced by a doubling of carbon dioxide?
Estimates vary by an order of magnitude, between less than 0.6° and 6°C.

The net warming produced by increased carbon dioxide does not equal just the increase in radiative forcing that is produced (3.6 watts/m^2 for a doubling), but is rather the sum of that forcing and all the feedback loops — which are incompletely known.

IPCC scientists, and their computer modellers, contend that the atmospheric warming caused by increased carbon dioxide will result in more water vapour in the atmosphere. This causes a strong positive feedback loop because the extra water vapour is itself a powerful greenhouse gas; adding its effect to that of the initial increase in CO_2 is an amplification of the initial forcing. Computer models that include this amplification predict a net warming of 3-6° C for a doubling of carbon dioxide.

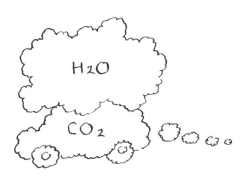

In contrast, independent scientists point out that the record of climate history teaches us that global climate is well buffered against change by both positive and negative feedbacks. Different methods of analysis have utilised either instrumental or historical proxy data, and also theoretical computer modelling, to estimate climate sensitivities that

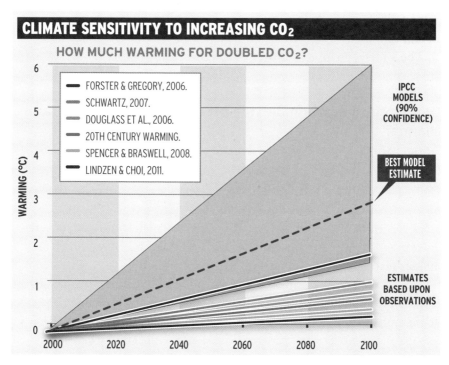

Fig. 17. Projected atmospheric warming that it is estimated will be caused by a doubling of carbon dioxide by 2100 (after Spencer, 2008). The climate sensitivity curves are plotted as projected by IPCC models (pink field), and (individual lines) as estimated by independent scientists based upon empirical measures. 20th century warming is also plotted (mauve line), for comparison.

range from about 0.3°C to 1.5°C for a doubling of carbon dioxide concentration.

IPCC projections of warming rely entirely on calculations from computer model representations of the climate system. These are claimed to incorporate all the physics of the climate system (Fig. 17) and to implicitly include both negative and positive feedback processes (V: But can computer models really predict future climate?). However, many of the important processes, including clouds and surface energy exchanges, operate on scales below that of the computer model representation. These processes therefore have to be approximated in the models by using assumptions and interpolations. There is no assurance that the various assumptions and approximations do not significantly amplify the climate model sensitivity

Several recent studies have measured the variation of radiation emitted to space, and shown that, at present levels of carbon dioxide, negative feedbacks outweigh positive feedbacks as temperature increases. This is

l empirical evidence that human-related carbon dioxide emissions
:ausing dangerous warming, and that the IPCC models are faulty.

₁ ₁₁ widely different estimates of climate sensitivity that can be found in
the scientific literature are summarised in Fig. 17 (p.105).

Do changes in carbon dioxide precede or follow temperature change?

Temperature change occurs before carbon dioxide change on all time scales.

One of our great research resources for studying past climate change is data gathered from Greenland and Antarctic ice cores. The bubbles of air preserved in the cores allow the synchronous measurement of ancient carbon dioxide and temperature magnitudes back to more than three-quarters of a million years ago.

Studies that apply this method to older (strongly compressed) ice are of rather low temporal resolution, with each sample point representing a period of time of up to 1,000 years or more. Early studies at such resolutions in the 1980s showed that the carbon dioxide and temperature curves matched with startling fidelity across several glacial-interglacial cycles, peak for peak and trough for trough. Not unreasonably, the conclusion was drawn that these results provided the 'smoking gun' evidence that carbon dioxide was indeed the driver for glacial (Milankovitch) scale climate change (III: What are Milankovitch variations?).

Since the 1990s, however, improved analytical and ice core sampling techniques have yielded results of higher resolution. These newer studies show that the apparently synchronously matched behaviour of temperature and carbon dioxide was incorrect; instead, a clear lead: lag relationship is apparent. Contrary to the expectation of many scientists, however, the relationship is that changes in temperature lead carbon dioxide changes by between several hundred and up to about a thousand years (Fig. 18, p.107). This, of course, is the opposite relationship to that required for carbon dioxide to be the cause and temperature change the effect, as misleadingly depicted in Mr. Al Gore's film, *An Inconvenient Truth*.

On much shorter time scales, other research has traced the seasonal changes in atmospheric carbon dioxide that occur as deciduous trees in the northern hemisphere wax and wane with the seasons, with concomitant rises in atmospheric carbon dioxide during the autumn and winter (as plants shed their leaves, and cease photosynthesising) and falls during the spring and summer (as regrowth occurs, and photosynthesis takes up carbon dioxide). This research shows that changes in temperature over the short, annual time-scale again precede the parallel change in carbon dioxide, this time by about five months.

A third line of research, on Pacific seabed cores, shows that deep ocean temperature change also leads atmospheric carbon dioxide change. Deep-sea temperatures warmed by about 2° C between 19,000 and 17,000 years ago, leading a parallel tropical surface ocean warming and rise in atmospheric carbon dioxide (by outgassing from the ocean) by about 1000 years.

Thus the evidence clearly indicates that atmospheric carbon dioxide level is not the primary driver of global temperature change over a wide range of time scales, but rather is itself a consequence of parallel but preceding temperature change.

The ocean reservoir of dissolved carbon dioxide (38,000 Gt) is almost fifty

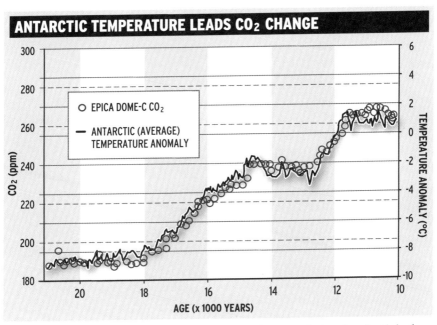

Fig. 18. Ice core studies have shown that changes in ancient atmospheric carbon dioxide level persistently lag parallel changes in temperature by up to 1,000 years. That temperature leads carbon dioxide, in this case by up to 200 years, is well demonstrated in a recent ice core study (Parrenin et al., 2013). Inflections in the average Antarctic temperature (black line) mostly precede changes in carbon dioxide concentration (red circles).

ιe magnitude of the amount in the atmosphere (780 Gt), and carbon is more soluble in cold than warm water. Therefore, the simplest explanation for the observed lead: lag relationship between temperature and carbon dioxide is that outgassing of oceanic carbon dioxide occurs at times of warming temperature, whereas extra oceanic carbon dioxide is dissolved in the ocean during cooling temperatures.[21]

How can the hypothesis of dangerous greenhouse warming be tested?
Against real-world data, and it repeatedly fails the test.

Climate science overall is complex. In contrast, the greenhouse hypothesis itself is straightforward and it is therefore relatively simple to test it, or its implications, against the available data.

The hypothesis that we wish to test is 'that dangerous global warming is being caused by man–made carbon dioxide emissions'. To be 'dangerous', at a minimum the change must exceed the magnitude or rate of warmings that are known to be associated with normal climatic variability.

Bearing these comments in mind, consider the following six tests:

- Over the last 16 years, average global temperature as measured by ground thermometers has displayed no statistically significant warming; over the same period, atmospheric carbon dioxide has increased by 8%. Large increases in carbon dioxide have not only failed to produce dangerous warming, but failed to produce any warming at all. *Hypothesis fails.*

- During the 20th century, an overall global warming of between 0.4°C and 0.8°C occurred, at a maximum rate, in the early decades of the century, of about 1.7°C/century (see also footnote 6, p.42). In comparison, our best regional climate records show that over the last 10,000 years natural climate cycling has resulted in temperature highs up to at least 2°C warmer than today (Fig. 5, p.29), at rates of warming and cooling of up to 2.5°C/century. In other words, both the rate and magnitude of 20th century warming falls well within the envelope of natural climate change. *Hypothesis fails, twice.*

- If global temperature is controlled primarily by atmospheric carbon dioxide levels, then changes in carbon dioxide should precede

[21.] Arguments against this simple interpretation can be advanced based upon detailed chemical isotope measurements of carbon. Readers interested in considering both sides of the arguments relevant to this are urged to consult some of the appropriate literature (see Recommended Reading, and papers referred to by Tom Segalstad here: *http://www.co₂web.info/esef5.htm).*

Explaining the pause in global warming despite increased CO_2 emissions gets harder for climate commissioner Tim Flannery. April 2013.

equivalent changes in temperature. In fact, the opposite relationship applies at all time scales (Fig. 18, p.107). *Hypothesis fails.*

- The best deterministic[22] computer models of the climate system, which factor in the effect of increasing carbon dioxide, project that warming should be occurring at a rate of at least +2.0°C/century, i.e. about 0.2°C/decade. In fact, no warming at all has occurred for more than the last decade, and the average rate of warming over the past 30 years of satellite records is little more than 0.1°C/decade. The models clearly exaggerate, and allocate too great a warming effect for the extra carbon dioxide (technically, they are said to overestimate the climate sensitivity) (compare Fig. 20, p.120). *Hypothesis fails.*

- The same computer models predict that a fingerprint of greenhouse-gas-induced warming is the creation of an atmospheric hot spot at heights of 8-10 km in equatorial regions, and enhanced warming also

[22] For an explanation of this term, see V: What are the main types of computer model?

near both poles. Direct measurements by both weather balloon radiosondes and satellite sensors show the absence of surface warming in Antarctica, and a complete absence of the predicted low latitude atmospheric hot spot (compare Fig. 22, p.129). *Hypothesis fails.*

One of the 20th century's greatest physicists, Richard Feynman, observed about science:

> *In general, we look for a new law by the following process. First we guess it. Then we compute the consequences of the guess to see what would be implied if this law that we guessed is right. Then we compare the result of the computation to nature, with experiment or experience; compare it directly with observation, to see if it works.*
>
> *It's that simple statement that is the key to science. It does not make any difference how beautiful your guess is. It does not make any difference how smart you are, who made the guess, or what his name is. If it disagrees with experiment it is wrong.*

None of the six tests above support or agree with the predictions implicit in the hypothesis of dangerous warming caused by man-made carbon dioxide emissions. Richard Feynman's description of scientific method is correct. Therefore, the dangerous AGW hypothesis is invalid, and that at least six times over.

Is atmospheric carbon dioxide a pollutant?
To term carbon dioxide a pollutant is an abuse of language, logic and science.

In western countries, including Australia and New Zealand, it is now widely believed that carbon dioxide is a dangerous pollutant whose level in the atmosphere needs to be controlled. This grotesque misconception did not

arise by accident, but is the result of a skilful propaganda campaign mounted since the early 1990s by environmental lobby groups and their media and political supporters. Controlling a debate by controlling the language is, of course, a key propaganda technique (II: Why all this talk about carbon instead of carbon dioxide?). The level of carbon dioxide in the atmosphere over the last 500 million years has varied between about 0.5% (5,000 ppm) and 0.03% (280 ppm) (below: Are modern carbon dioxide levels unusually high, or dangerous?).

This cartoon appeared the day that John Brumby's Victorian state government was defeated at the polls. Despite Hazelwood power station's reputation of being the dirtiest because of its CO_2 emissions, it has a clean pollution rating from the EPA. November 2010.

Because it is a greenhouse gas, more carbon dioxide in the atmosphere, other things being equal, does cause warming (above: What is a greenhouse gas?).

But other things are far from equal, two important considerations being, first, that the extra warming diminishes in magnitude rapidly (logarithmically) as carbon dioxide increases (Fig. 16, p.103); and, second, that many and varied feedback loops exist in the natural world which act in some cases to enhance and in other cases to diminish the amount of extra warming that occurs (above: What net warming will be produced by a doubling of carbon dioxide?).

These facts indicate that negative feedbacks (i.e. cooling) probably dominate over carbon dioxide-forced warming over the range of geologically usual levels of carbon dioxide concentration. One such negative feedback that many scientists think is important is an increase in low level clouds, which, by reflecting incoming solar radiation back to space, causes cooling.

The points considered so far are concerned with the physical effects of carbon dioxide. But the molecule is also the key for one of the most critical biological functions on our planet. For by furnishing plants with the essential material that they need to photosynthesise, carbon dioxide underpins all plant growth; it is, in effect, plant food.

To the degree that presently increasing concentrations of carbon dioxide might cause mild warming — and noting that our planet is currently traversing a short warm interval in an extended series of glaciations — more carbon dioxide is likely to be beneficial. Where plant growth is concerned, however, 'likely' has nothing to do with it, for it is certain that moderate increases in carbon dioxide beyond present levels (say to a doubling or tripling) will enhance plant productivity; combined with which, plants use water more efficiently at higher carbon dioxide levels. Recent studies have estimated that between 1989 and 2009 about 300,000 km^2 of new vegetation became established across the African Sahel region in parallel with the increasing levels of atmospheric carbon dioxide.

In other words, the recent increases in carbon dioxide have have helped to green the planet and feed the world. What's not to like?

Are modern carbon dioxide levels unusually high, or dangerous?

No. Compared with geological history, Earth currently suffers from carbon dioxide starvation.

Modern measurements

Accurate chemical methods of measuring the amount of carbon dioxide in the atmosphere were well established by the mid-19th century. Over the next hundred years, hundreds of thousands of such measurements suggested that the background level of carbon dioxide in the atmosphere was about 280-300 ppm, though some measurements reached 500 ppm or more. Great variability in concentration was observed in accordance with both daily and seasonal cycles, and with geography; for example, northern hemisphere measurements made in winter are higher than those made in summer

(because deciduous trees cease photosynthesis, and thus carbon dioxide uptake, in winter). Enhanced levels have also been found in samples collected from the vicinity of industrial plants or power stations.

From 1958 onwards, a continuous series of modern chemical measurements has been made by scientists of Scripps Institute of Oceanography at a site near the summit of the Mauna Loa volcano in Hawaii. Because the site is high in the atmosphere, and located in the middle of a large ocean, the results are assumed to reflect the genuine background concentration of the well mixed atmosphere. Nevertheless, a significant proportion of the observations are discarded because of interference from local sources and sinks. The Mauna Loa record exhibits well the strong seasonal signal caused by northern hemisphere deciduous plant growth and hibernation, superimposed on a background curve that climbs steadily from 360 ppm in 1958 to 390 in 2011 (Fig. 7, lower, p.36).

Assuming that most of the increase from 280 ppm of carbon dioxide in the early 19th century to 390 ppm a little over a hundred years later has been caused by human-related emissions, the key questions are whether such an increase is unusual, or anything to worry about. To answer these questions requires that we look at older, geological records of carbon dioxide levels.

Ice core measurements

Bubbles of contemporary air are trapped in the layers of snow that are compressed with time to turn into ice layers in the Greenland and Antarctic ice caps. Since the 1980s, scientists have been able to extract samples of ancient atmospheric gases (air) from ice-core samples, and thus directly determine carbon dioxide levels back to almost a million years ago. Measurement of these ancient air samples has confirmed that over the recent geological past atmospheric carbon dioxide levels have fluctuated broadly in sympathy with the major glacial and interglacial climatic episodes, typically being 280 ppm during warm interglacials but plunging to a dangerously low 180 ppm during the cold glaciations; 'dangerously' because at 150 ppm most plants cease to function adequately and at such a level a global biodiversity crisis would undoubtedly occur.

Thus both the ice core and the historic chemical measurements agree that the pre-modern industrial revolution level of carbon dioxide was about 280 ppm.

Older geological measurements

However, it is important to know also how the carbon dioxide concentrations found today and in the ice core samples compare with levels during the deeper geological past. Reconstructing deep time atmospheric carbon dioxide is difficult because of the lack of ancient air samples older than the ice cores. Accordingly, geochemists use a number of indirect chemical measurements, such as analysing carbonate soil nodules which, having grown in chemical equilibrium with the atmosphere, contain a stored signal of atmospheric carbon dioxide concentration.

The results indicate (Fig. 19, p.115) that in the recent geological past the Earth has been in a state of carbon dioxide starvation compared with most of the previous 500 million years. And this is still true today, even after human sources have helped to add up to 100 ppm of carbon dioxide to the atmosphere.

From high levels of about 5,000 ppm in the Cambrian Period (500 million years ago), carbon dioxide decreased steadily to about 1,000 ppm between the Devonian and Carboniferous Periods (450-350 million years ago). Since then, though with fluctuations up to 2,000 ppm, the average level has decreased further to the pre-industrial low level of 280 ppm. While the reasons for many of the smaller-scale fluctuations in this record remain unknown or controversial, the initial draw down coincided with the evolution and colonisation success of land plants, as manifest in the extensive and thick post-Devonian coal deposits that occur on all continents, including Antarctica. Starting a little later, ocean phytoplankton too helped to remove carbon dioxide from the atmosphere, as represented in the sedimentary record by the chert (SiO_2) and chalk $(CaCO_3)$ deposits that are made up of the tiny skeletal remains of species of phytoplankton that possessed hard-part skeletons.

Thus when industrial nations dig up and burn coal and other fossil fuels to generate electricity, and for other industrial and community needs, the waste carbon dioxide gas is returned to the atmosphere — from whence it came. It is hard to understand why this is perceived to be a problem.

Conclusion

In summary, modern carbon dioxide levels are exceptionally low, as testified by the globally widespread coal deposits that represent the storage sink for carbon dioxide that has been extracted from the atmosphere by land plants through the ages. Returning that carbon dioxide to the atmosphere is environmentally beneficial, because it helps to green the planet and feed the world.

What about methane, then?
Methane, present in the atmosphere in only parts per billion, is but a minor greenhouse player.

Methane (CH_4) is a naturally occurring greenhouse gas that forms from the decay of biological material in the absence of air. Sometimes called marsh gas, it is found in association with wetlands and irrigated crops, especially rice. Methane is also the major component of natural gas, and escapes to the atmosphere from leaking pipelines and storage depots. Methane breaks down naturally in the atmosphere to form carbon dioxide and water vapour, and has a short lifetime of only about 10 years.

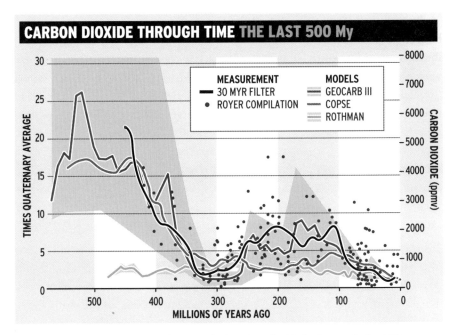

Fig. 19. Reconstructed proxies for atmospheric carbon dioxide levels over the last 550 years, the period during which multi-cellular organisms originated and diversified. The overall trend is one of diminishing carbon dioxide, with a steep drop starting about 450 million years ago at the time of origin of photosynthesizing land plants. Dark blue line, average from 372 measurements of palaeo-atmospheric proxies; orange-red line and shaded pink error zone, projections from the GEOCARB III model.

Methane is present in the atmosphere at very low concentrations and is measured in parts per billion (ppb). Paleoclimate records indicate that methane concentration was higher during earlier geological times when Earth was wetter. In recent historic and modern times, methane concentration has increased from about 700 ppb in the 18th century to the current level of near 1800 ppb today. The increase in methane concentration levelled off and stabilised between 1998 and 2006 at around 1750 ppb, which may reflect measures that were taken at that time to stem leakage from wells, pipelines and distribution facilities. More recently, however, from about 2007, methane concentrations have started to increase again. One possible explanation for this is that it may reflect new leakages associated with the introduction of widespread fracking[23] in pursuit of unconventional oil and gas.

[23.] Fracking is shorthand for the 'rock fracturing' process that is used in the retrieval of oil and gas from unconventional reservoir rocks such as shale or coal. It is accomplished by the pressurised pumping of a mixture of water, dilute chemicals and sand down a well, in order that the fluid penetrates and widens existing natural cracks and crevices in the rocks, thus allowing contained hydrocarbons to flow out.

terms of atomic weight, methane is a 7-times more powerful green-gas than carbon dioxide, but it is present in such low concentrations that it contributes relatively little to the overall greenhouse effect. The contribution of increased methane to radiation forcing since the 18th century is estimated to be about 0.7 W/m², which is small. Because methane is a valuable fossil fuel, commercial imperatives are a constant constraint on unnecessary leaks to the atmosphere. *No mention of the methane supposedly escaping from thawing tundra.*

Do I have to worry about ozone too?

Ozone is a greenhouse gas, but like methane it is only a bit player in terms of global warming.

Ozone (O_3) is a short-lived greenhouse gas that forms by different processes in both the high and low atmosphere. Stratospheric ozone in concentrations of 500-1000 ppb is located between 20 km and 40 km altitude, and forms through the dissociation of oxygen molecules by incoming ultraviolet radiation from the Sun. The free oxygen atoms then associate with other oxygen molecules to form ozone. Ozone also forms in the lower atmosphere during electrical storms and as a consequence of urban pollution, especially vehicle exhaust fumes (up to 15 ppb in smog).

In the high atmosphere ozone is very beneficial, because it protects life at the surface from harmful ultraviolet radiation from the Sun. However, in the low atmosphere ozone is corrosive, even in small concentrations. Ozone is relatively short-lived and its low concentration in the lower atmosphere is now semi-permanent because of the prevention of photo-chemical smog by urban air pollution control measures.

As a greenhouse gas, a molecule of ozone is estimated to have 2,000 times

the warming effect of a molecule of carbon dioxide, but its low and variable concentration in the low atmosphere means that any enhancement of the greenhouse effect is small and transitory. Regulation in order to avoid photo-chemical smog and ozone formation is aimed at health impacts and the prevention of corrosion on plants and materials. Such measures have the additional notional, and at the same time arguable, benefit of preventing the small incremental greenhouse effect that the extra ozone would otherwise cause.

V

COMPUTER MODELLING

What is a climate model?
A theoretical representation of part or all of the climate system.

The climate system is very complex, and made up of many interacting components and processes. In order to better understand it, scientists construct models that summarise those parts of the system that they wish to study. In all cases, it is the Earth's actual, observed climate system that is the basis for the model. Unless there are measurements and knowledge of the interacting climate processes that occur in the real world, then it is not possible to construct a meaningful model.

Understanding of the Earth's climate system has evolved over the last 100 years as systematic observations and analysis (called empirical evidence) have improved our knowledge. Essential laws of physics have been discovered that can quantify some of the major processes that regulate the climate system, enabling the representation of climate in terms of a complex web of interacting components and processes. Though many general physical laws, including the Gas Laws, Laws of Thermodynamics and the laws relating to electromagnetic radiation were developed in isolation from climate science, nevertheless climate models must conform to these laws.

Hell freezing over. July 2008.

Many of the empirical relationships used by scientists in description of the climate system have evolved without full understanding of all underlying and linking processes. Accordingly, the theoretical understanding of the climate system has evolved with time, as initial theoretical constructs have come to be significantly modified or overturned after new observations came to hand. For example, balloon-borne instruments gave an understanding of the temperature profile of the atmosphere, but it was not until high altitude aviation became common during World War II that the existence and importance of strong jet stream winds became apparent.

However, some important components of the climate system remain comparatively undocumented to this day. Logs from early sailing ships, and more recent systematic observations, provide good knowledge of the surface ocean currents, but it is only in the last decade that Argo buoys have provided systematic data about subsurface currents and temperatures, and how these vary with time. The scale of many currents, including the nature of eddies, has come as a surprise.

Models are constructed from known empirical relationships. If a model can simulate a past climate change with fidelity, say a global temperature warming or cooling, then the possibility has been created to use the model to project a future state. However, the ability to reproduce a change from one past state to another does not, on its own, guarantee that a model will accurately project a future state. Without complete knowledge of the actual climate system, all models involve assumptions and approximations; no matter how carefully models are constructed to match the knowledge of past events, it cannot be guaranteed that they will accurately project future events.[24]

The projection of future climate states is particularly hazardous because of the relatively short record of instrumental measurements available with which to document and model the climate system. The lack of data for important components of the system such as deep ocean circulation, and our incomplete knowledge of important processes such as clouds and aerosols, add greatly to the uncertainty, especially where deterministic modelling is concerned.

What are the main types of computer model?
Two main types: deterministic and statistical-empirical.

Deterministic models

The computer models that the IPCC uses to project future climate are called General Circulation Models, or GCMs for short. These models are deterministic, in reference to the fact that mathematical equations form their kernel, so that, from a given set of starting conditions, the model calculates successive transitions from one state of climate to the next forward though time. GCMs can also be used to help assess the relative importance of various climate process interactions, and the role of external forcing as against internal variability in producing particular changes. The IPCC draws on more than 30 such GCMs (including one maintained by CSIRO), which are run by major climate research groups around the world (Fig. 20, p.120).

Because Earth's changing climate is complex, GCMs need to be based upon extensive knowledge of matters as diverse as normal weather process-es, solar radiation, cosmic radiation and its influence on cloudiness, slowly varying ocean currents and, on very long time scales, even the drifting of continents (I: What is climate?).

It is, however, an iron fact that we do not possess a complete knowledge of many important climate processes, for example an accurate quantitative understanding of the influence of aerosols or clouds. The climate system therefore cannot be fully specified mathematically within a GCM, and

---

[24.] Note that it is not correct to say of current climate models that they predict future climate, despite most reporters and politicians believing that to be the case. In science and engineering, the word predict is only used in the context of calculations that are known to possess a predictive or forecast skill within specified conditions. Climate models have never been validated and therefore have no specific predictive ability.

uncertain processes have to be included as 'best guess' approximations that are called parameterisations. All GCMs contain many parameterisations, and slight variations in the numerical value allotted to them can greatly influence the outcome of any particular model projection — including the magnitude of any warming that might result from systematic increases in atmospheric carbon dioxide. The output of any model run, then, is a 'what-if' thought experiment, with the 'what' being the magnitude of change in global temperature, and the 'if' referring to the characteristics of the model, including imposed forcings such as an increasing carbon dioxide concentration.

By suitably adjusting various parameters, GCMs have been able to repro-

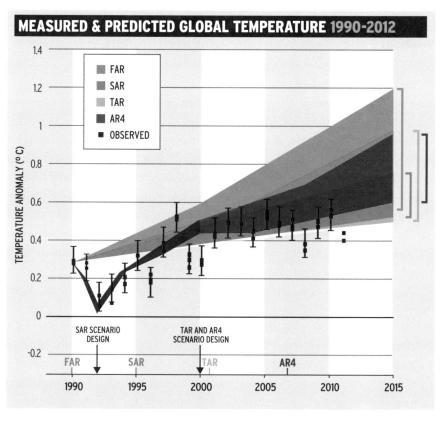

Fig. 20. Comparison of measured and modelled global temperature since 1990, as presented in the draft of the IPCC's 5th report, Fig. 1.4 (Cubasch & Wuebbles et al., 2012). The vertical groups of black data squares for each year represent NASA GISS, NOAA and UK Hadley Centre estimates of the global temperature from averaged observations (whiskers, 90% uncertainty range). Coloured shading indicates the overlapping, model-projected ranges of annual global surface temperature from 1990 to 2015 for models used in previous IPCC reports. All values are aligned to match the observed average value in 1990. Note that the model projections are persistently too warm in comparison with the observed data values.

duce many features of the global climate system, and to match the broad shape of the global temperature curve over the last century. This latter characteristic is claimed to give confidence in the models, but the reality is that no model has yet been shown to possess genuine predictive skill when applied to independent data sets.

Since the first IPCC report of 1990, their GCMs have consistently project-ed a rate of increase of global temperature due to increasing carbon dioxide that is between 0.2°C and 0.3°C per decade, with an uncertainty range of 0.2°C to 0.5°C/decade. In reality, however, global temperature has only risen about 0.2°C since 1990 (or 0.1°C/decade). This is about half the lowest value of the IPCC's projected range, with no rise at all having occurred over the most recent decade (Fig. 20, p.120).

Despite their complexity, current GCMs remain rudimentary in their construction when compared to the reality of the Earth's dynamic climate system. There are significant limitations on their representation of important physical processes that occur in the atmosphere and ocean. Specifically:

- The models do not adequately represent the full range of energy ex-change processes that occur naturally; particular deficiencies exist in the way that the models treat clouds, the transfer of heat and moisture between the Earth's surface and atmosphere (including precipitation), and the growth and decay of ice sheets;

- Only limited observations exist of sub-surface ocean circulations and their variability. Better description and understanding of these ocean circulations is a crucial prerequisite for the construction of accurate models of the natural variability of the climate system;

- Current models do not take into account, and therefore cannot predict, major and important short-term climate-moderating mecha-nisms, such as El Niño; nor intermediate length multi-decadal climate oscillations such as the Pacific Decadal Oscillation; nor longer time-scale climate episodes such as the occurrence of the next Little Ice Age or the start of the next major glaciation; and

- Current models do not take into account several important solar factors, including especially variations in the Sun's magnetic field and the way that such changes impact on Earth's weather — including in-directly through possible effects exercised on cosmic ray penetration.

A common claim for the climate models used by the IPCC is that their ability to replicate (hindcast) the global temperature record of the 20th century represents their validation. This is untrue, as the replication might instead indicate skilful curve matching. For confident projection to a future state, a model must be constructed and tuned on one empirical data set, and then be validated by projection over a separate and independent data set. This has not been accomplished for any of the IPCC's deterministic models.

Nonetheless, given time two factors will improve the utility of climate models to project future climate states. First, the isolation of component parts of a model for detailed study can identify those parameterisations to which the models are sensitive, leading to more focussed research and a reduction in uncertainty. Second, the accumulation of longer records of measured climate data will eventually generate sufficiently long historical records for the models to be better tuned and validated; analysis of validation errors will then identify parameterisations which require refinement, again leading to reduced uncertainty.

In summary, confidence in the projections made by the current generation of deterministic climate models is low because their construction is based on only a short period of climate history, and because they have not been validated on independent data. For the moment, therefore, deterministic GCMs represent a highly constrained and simplified version of Earth's complex and chaotic climate system. Their value is heuristic[25], not predictive.

Statistical-empirical models

An alternative way to project future climate scenarios is to examine historical climate data, such as a temperature record, to see what pattern and underlying relationships may be contained in the data.

Many climate records exhibit decadal and multi-decadal cycles of behaviour. Such patterns can often be represented by an empirical[26] mathematical equation, and by extending that equation forward in time, a projection can be made of future climatic conditions. Note, importantly, that such modelling involves no claim that scientists understand quantitatively how the climate system works, but is based instead on simple statistical analysis plus the assumption that any pattern of change present in historical data will continue unchanged into the future.

Statistical-empirical models of climate have been fashioned using a wide variety of data, including analyses of past temperature and past solar behaviour over different time periods. For example, Nicola Scafetta has developed a

---

[25]. Heuristic — serving to discover; aiding understanding
[26]. Empirical model: one based solely on, and fitted to, real world data

model that uses harmonic analysis of the observed 9, 10, 20 and 60 year climate cycles (Fig. 21, black curve), and compares this with the range of IPCC computer projections (Fig. 21, green field). There is an excellent fit between the observations and the harmonic model for 1850-2010, and Scafetta's model predicts much less warming for 2010-2100 than do the IPCC computer projections. The harmonic model forecast after 2010 (Fig. 21, blue field with central black curve) assumes an estimated anthropogenic forcing of between 0.5 and 1.3°/century. This is unusual, in that most empirical models are based solely on natural climate variability, and its short term projection, and make no attempt to include possible anthropogenic effects.

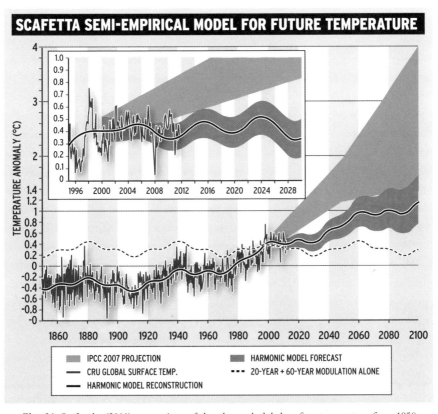

Fig. 21. Scafetta's (2011) comparison of the observed global surface temperature from 1950 to 2050 (red curve) fitted with an empirical model curve that is based on the known 9, 10, 20 and 60 year climate cycles (black curve); the range of IPCC computer projections is delineated by the green field. The model forecast after 2010 (blue field with central black line) assumes an estimated anthropogenic forcing of between 0.5 and 1.3°/century; the dotted blue line below projects the model result without this forcing. Inset: detail of same graph for the period 1996-2030, emphasizing that all IPCC models run too warm even when an anthropogenic forcing is included in the empirical model (cf. Fig. 20).

Nearly all published empirical models except Scafetta's (because of its inclusion of an anthropogenic effect) project that significant global cooling will occur over the first few decades of the 21st century. Interestingly, observational records to date do indeed point to the cessation of the mild warming seen in the late 20th century, and its replacement by temperature standstill or cooling over coming decades (Fig. 1, p.17).

Why do we need computer models to study climate change?
Because models provide quantitative projections of future climate.

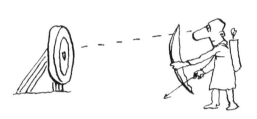

We live in a managed society that is kept running in manifold ways by the analysis and use of rigorous, quantitative data. Whether it be planning the food supplies needed by a city, planning the irrigation networks for a country district, or planning state and national transport systems, the assembly of a set of accurate facts about the past and present behaviour of a system is an essential prelude to making quantitative projections about likely future needs. The availability and use of computers greatly aids the calculation of such projections.

With respect to climate, perhaps the most important consideration is the impact that extreme events have on society, which in Australia routinely involves the impacts of cyclones, floods, droughts and bushfires. Understanding the climate variability that leads to such events is essential to provide for societal resilience, and to minimise the death, destruction and community disturbance that can accompany such events. Civil defense efforts to counter such hazards are traditionally based upon the systematic recording and statistical analysis of past climate data, in order to define an envelope of hazard risk, size and recurrence interval for planning the response to future events.

The 1985 UN Villach Conference made the bold statement that past data are no longer sufficient for planning, because increasing human-caused carbon dioxide in the atmosphere and global warming are changing the climate system in a fundamentally new way; future planning therefore must be on the basis of deterministic computer model projections. But the extreme complexity of the climatic system and its internal processes makes it impossible to capture them accurately in a computer model (above: What are the main types of climate model?). Nevertheless, try we must, because it is only by delineating different 'what-if' experiments about possible future outcomes that emergency service managers can be best advised about how to plan for a likely range of future climate hazards.

The great value of deterministic modelling in a suitable context is well exemplified by the now routine accuracy of weather forecasts up to several days into the future. Climate GCMs have evolved from these models, and it is therefore often suggested that the success of the weather forecast models to several days ahead should give confidence in GCM projections over the longer term. But this is an invalid comparison. For although the successful weather forecast models are closely similar to the atmosphere component of models used to attempt climate forecasting, the chaotic nature of changing weather patterns precludes (and may always preclude) accurate prediction beyond the week-long horizon of weather forecasts. As IPCC scientists themselves concluded in 2001 (IPCC 3rd Assessment Report, p.774):

> *In climate research and modelling, we should recognise that we are dealing with a coupled non-linear chaotic system, and therefore that long-term prediction of future climate states is not possible.*

Despite their lack of predictive skill, however, the use of GCMs does confer an important benefit because the process of modelling is a heuristic one. By slightly changing various parameters in a model, and observing how the output then changes (or not) in turn, it may one day be possible to learn more about the sensitivity of climate to particular influences and about the range of possible climatic responses to changing circumstances.

But can computer models really predict future climate?
No — at least, not GCMs.

In the usage of engineers — who, remember, carry the responsibility for ensuring that the bridges of the world don't collapse — a computer model that is to be used for real-world design purposes must first have been validated.

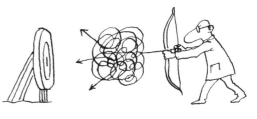

Validation involves rigorous testing that demonstrates the capability of the model to forecast the future behaviour of the modelled system for a range of conditions and to a specified and acceptable level of accuracy. Generally validation is against completely independent empirical data. No such procedure has been carried out for any of the GCM models deployed by the IPCC.

er, the models are claimed to be accurate based on their ability to reproduce the 20th century temperature record. However, it is this temperature record that has been used in the construction of the various models in the first place!

The computer models used by the IPCC therefore remain unvalidated because they have not been tested against independent data; as a consequence, the degree of skill inherent in GCMs is unknown. It is because the models have not been validated that the outputs of GCMs are called 'projections' and not 'predictions' or 'forecasts'. The outcome of a GCM run is a scenario; it describes one outcome or state within an envelope of manifold alternative possible outcomes or states. In the event, too, comparison of model projections with recent climate trends suggests that the models currently exaggerate the influence of human-caused carbon dioxide emissions.

Regrettably, the vital distinction between prediction and projection regularly escapes politicians and media commentators, and thereby there is a widespread, and wrong, belief amongst the general public that deterministic climate models provide accurate climate forecasts or predictions.

The game is rather given away by the following style of disclaimer, which is inserted inside the front cover of all climate consultancy reports prepared by CSIRO, one of the groups that acts as a model-provider to the IPCC:

> *This report relates to climate change scenarios based on computer modelling. Models involve simplifications of the real processes that are not fully understood. Accordingly, no responsibility will be accepted by CSIRO or the QLD government for the accuracy of forecasts or pre- dictions inferred from this report or for any person's interpretations, deductions, conclusions or actions in reliance on this report.*

If GCMs are unable to produce predictive outputs then by definition they are unsuitable for direct application in policy making. It is clearly inappropriate to use projections from unskilled computer models for planning purposes as if they were confident, validated predictions of future climate. Instead, future environmental planning needs to be based upon careful analysis of the relevant real world, empirical data.

Do computer models suggest 'fingerprints' for human-caused warming?
Yes, but they are not necessarily unique, nor always present.

Deterministic computer models are based upon our current understanding of the fundamental physics of the climate system. They therefore should be able to make predictions about how global climate might change in response to increases in man-made carbon dioxide emissions. Such change includes the possibility that new characteristics that were not part of the natural, pre-industrial climate system may emerge in a projected future climate state.

Misleadingly, the possible new climate characteristics have come to be called 'fingerprints' of man-made warming. Though applicable to a prediction that is a unique result of a theoretical increase in carbon dioxide, the term fingerprint is not appropriate to describe model outcomes that are able to be explained by more than one cause — which is the case for nearly all the outcomes on the list below.

Modelling groups claim that the following outcomes will result from an increase in man-made carbon dioxide emissions, and that these outcomes are observed as 'fingerprints' in the models:

1. Shallow ocean temperature will increase
2. Surface atmospheric temperature will increase
3. There will be more extra warming at night than during the day
4. The middle atmosphere (stratosphere) will cool
5. The upper atmosphere (thermosphere + ionosphere) will cool
6. The rate of temperature increase will be higher at 8-10 km in the lower atmosphere (troposphere[27]) than it is at ground level
7. Atmospheric pressure at sea-level will increase over the subtropics and decrease in polar regions
8. Rainfall will vary across the warmer parts of the two hemispheres, decreasing in the northern hemisphere and increasing in the southern hemisphere
9. An increase will occur in downward longwave (infrared) radiation
10. A decrease will occur in upward longwave (infrared) radiation
11. The changes in longwave radiation will produce an energy imbalance of $0.9\,W/m^2$ at the top of the atmosphere

There are three major problems with this list of claimed predictions. The first is that none of these outcomes uniquely requires that the greenhouse forcing claimed to cause these changes is a result of human (as opposed to

[27.] The troposphere is the term given to the lowest, humid part of the atmosphere, within which major weather systems and turbulent convection occur (Fig. 3, p.21). Temperature decreases rapidly with height in the troposphere, at a rate of about 6.5°/km. Jet airliners generally fly at 10-12 km height in the upper troposphere, the top of which, and boundary with the overlying stratosphere, occurs at a height of about 20 km in the tropics, decreasing gradually to about 7 km in polar regions.

natural) increases in emissions, i.e. even should they be met, the listed criteria are not fingerprints of human causation but at best hints or circumstantial evidence.

The second problem is that these changes are not uniquely caused by greenhouse emissions, but can all have other causes. For example, increases in solar magnetic and ionic activity have a direct effect on the warming and cooling of the middle and upper atmosphere, through modulating the amount of ozone present. Or, to take another example, an increase in upward radiation might result from an increase in the reflectivity (albedo) of the Earth's surface.

Though outcomes 9 and 10 do uniquely require the presence of extra greenhouse forcing, these outcomes apply whether or not the extra greenhouses gases stem from human sources; they also apply whether or not the increase is accompanied by global temperature change.

The third problem is that although atmospheric measurements in some cases record that changes similar to the model projections listed above have occurred, in other cases the measurements flatly contradict the modelled outcomes.

A notorious example is the prediction of a faster rate of warming in the upper troposphere to produce a so-called 'greenhouse hot-spot'; models almost unanimously flag this to be a signature of human emissions. In the real world, however, neither the radiosonde measurements of temperature available since 1958, nor the satellite-mounted measurements available since 1979, show any indication at all of the presence of such a hot spot. Instead, the measurements show quite clearly that faster rates of warming apply near the ground than in the upper troposphere (Fig. 22, p.129)

Though computer model projections do frequently exhibit new, changed or enhanced characteristics of the climate system, none of these can unequivocally be tied to human-caused global warming. In some cases they reflect the expected response of a warming Earth whatever the cause of the warming; in others they are unique to a particular GCM, and therefore not necessarily representative of the climate system. Describing them as 'fingerprints' is therefore inappropriate.

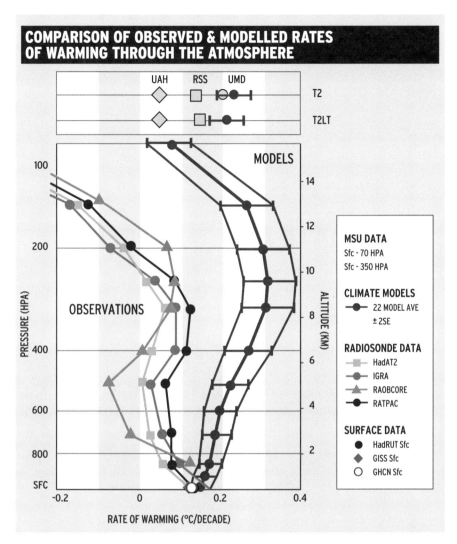

Fig. 22. Comparison between model predictions (red points and lines) and actual measurements of the rate of temperature change (blue, orange and green points and lines) plotted against height in the atmosphere (Douglass et al., 2007). Vertical scales are atmospheric pressure (left) and equivalent atmospheric height (right); horizontal scale depicts the rate of warming in degrees C/decade. All observations indicate a lesser rate of warming than do the model projections, and the 'hot spot' indicated by the models at about 10 km height in the atmosphere is absent from the data plot.

VI

CLIMATE AND THE OCEAN

Can we take the temperature of the ocean?
Yes, in principle, but only for about the last two decades with accuracy.

The relative densities of air and water mean that the total mass of the atmosphere is equivalent to that of only the top 10 metres of the ocean. Correspondingly, the heat contained in the atmosphere is equivalent to that contained in only the top 3.2 metres of the ocean. Oceans being, on average, about 4,000 metres deep, they obviously comprise a very large heat reservoir.

It is for these reasons that the oceans are sometimes called the thermal and inertial flywheels of the climate system. The oceans confer a relative stability on the climate system, and when changes in ocean characteristics occur, such as through an ENSO cycle, this quickly and markedly affects the atmosphere and its weather systems.

Given these facts, many scientists argue that the graph of historic atmospheric temperature (Fig. 1, p.17) is less suitable as a means of judging global climate change than would be an equivalent graph of ocean temperature, or, even better, a graph of ocean heat content.

A record of changing ocean surface temperature over the last 100 years exists, and displays broad similarities in shape with the air temperature curve (i.e., shows 20th century warming). However, early surface water temperatu measurements were made from sailing ships using samples collected with different types of containers — including metal, wooden, plastic and canvas buckets, all of which exhibit different thermal behaviour. As sail gave way to steam, engine inlet water was used to measure 'surface' temperatures, but ships of different size have inlets at different intake depths, which vary again according to the amount of cargo on board. In addition, few of the measurements used to compile the temperature graph were systematically recorded from exactly the same location over time. Accordingly, the observations on which the ocean surface temperature graph is founded are of doubtful accuracy prior to about the last two decades, at which time accurate measurements made from drifting buoys and satellite sensors became available (Fig. 23, upper, p.132).

For the past half century, marine temperature observations have also been collected from the subsurface ocean under an international programme coordinated by the World Meteorological Organisation. Selected regular commercial ships deploy expendable bathythermograph (XBT)[28] instruments to register temperature and salinity at depth. Meanwhile, a determination grew in the scientific community to deploy a worldwide system of temperature-measuring buoys that would gather accurate measurements throughout the upper ocean. The first buoys of what is known as the Argo network were launched in 2003, and more than 3,000 such buoys now operate throughout the world ocean. Each buoy descends to a depth of 2,000 metres, after which it measures the temperature through the water column as it ascends slowly to the surface, where it radios the results back to the laboratory via satellite.

It was widely anticipated that the results of the Argo system would show a warming trend in the ocean of 2°C/century, as predicted by climate models. In reality, nine years of observations now demonstrate that the trend in ocean heat content has been either flat or slightly declining during the period since 2003 (Fig. 23, lower).

Though it is early days yet, this lack of warming in the Argo measurements mirrors the atmospheric temperature measurements made over the same period (compare Fig. 9, lower, p.75). Taken together, then, the ocean and atmospheric measurements establish that the late 20th century phase of global warming has stopped, a fact conceded in official releases from the British Meteorological Office in October 2012 and January 2013, and also by Rajendra Pachauri,

---

[28.] An expendable bathythermograph consists of a small instrument probe attached to a spooled wire; as the probe drops through the ocean it transmits changing temperature and salinity measurements back through the wire to a shipboard data-logger. XBTs have an assumed rather than measured rate of descent though the water column, and any actual departure from the presumed standard rate of fall introduces error.

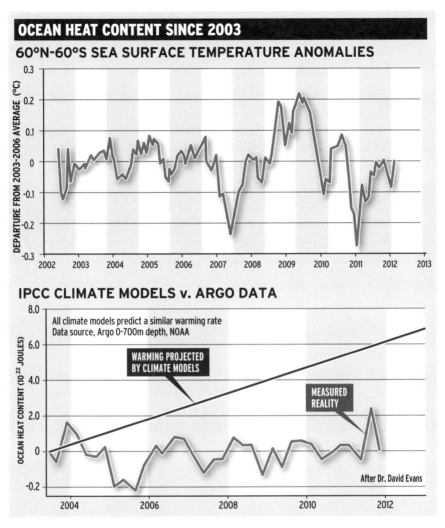

Fig. 23. Accurate measurements of global ocean heat content only became available with the deployment of the Argo system of ocean buoys in 2002-2003. These graphs show the lack of any significant increase in either (upper) ocean surface temperature (NASA, RSS; Spencer, 2013), or (lower) ocean heat content over the last ten years (Argo; Evans, 2013). This lack of warming parallels the lack of recent warming seen in the atmosphere (Figs. 1, 9).

Chairman of the IPCC, in February, 2013.

An important effect of a heating ocean is the expansion that occurs in the warmer water, which drives a global sea-level rise. Thus the slackening of the rate of sea-level rise that has been recorded over the last few years by both satellite radar ranging and tide gauge measurements is also consistent with the lack of warming of the ocean.

No increase in air temperature, no increase in ocean temperature and no increase in the rate of global sea-level rise. These three independent but supporting indicators suggest that it is now past time to rethink climate policies that are aimed at 'preventing' dangerous global warming.

Why is it important to distinguish between local and global sea-level change?

Because LOCAL relative sea-level change is relevant to coastal planning.

Sea-level rise is one of the most feared impacts of any future global warming. But public discussion of the problem is beset by poor data and extremely misleading analysis, which leads to unnecessary alarm. A proper understanding of the risks associated with sea-level change can only be attained by maintaining a clear distinction between changes in global sea-level (often also called eustatic sea-level) and changes in local relative sea-level.

Change in local relative sea-level is commonplace along coasts, and has been observed by mankind for millennia. A rising sea-level can adversely impact on local communities, and even give rise to legends, like that of Atlantis. As well, and even at the same time as sea-level is rising in one place, a fall in relative sea-level can occur in other places. That this is so relates to the fact that tectonic movements of the Earth's crust differ from place to place, in some places sinking and in others rising. Coast lines are also eroded landwards by wave action, and elsewhere extended seawards as river deltas grow through sediment deposition.

In many places, historically, appropriate human engineering responses have been made to changing sea-level. For example, sea-level rise has been combatted by the building of sea walls and protective dykes. In applying such measures, the Dutch have sensibly based their coastal planning on scientific knowledge of locally observed rates of shoreline and sea-level change.

Global sea-level change

Global sea-level change corresponds to differences in a notional worldwide average sea-level. The statistic broadly corresponds to the volume of water contained in the oceans. Changes can be brought about by either global warming, which expands the ocean volume and also adds water through the melting of land ice, or global cooling, which acts in a converse manner.

The best data from tide gauge measurements indicates that average global sea-level has been rising at a rate of 1.7 mm/year for the last 100 years. Though such calculations and projections can be made as an average, the results have little use for coastal management in specific places because local tectonic movements or dynamic oceanographic and sedimentary factors may dominate locally over the global sea-level change factor.

Local sea–level change

In contrast, local relative sea-level can be and is measured at specific coastal locations, and it is affected by the movement up or down of the geological

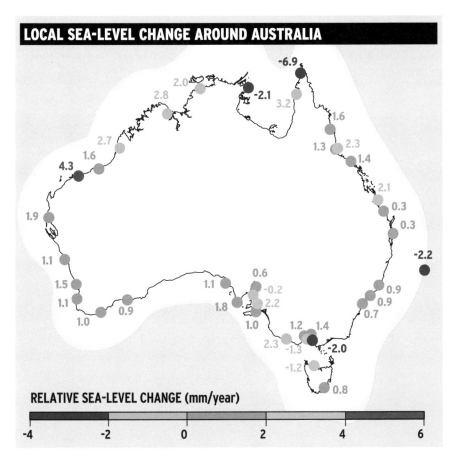

Fig. 24. Rates of local relative sea-level change in mm/year for locations around the Australian coast for which tide gauge records of at least 25 years in length are available (BOM, 2009). Note the wide range of variation between sea-level falling at rates up to -6.9 mm/year adjacent to Torres Strait and rising at rates up to +4.3 mm/year elsewhere. Around most of the continent, local sea-level is rising at around 1 mm/year, which is only one-third of the current estimates that global sea-level is rising at more than 3 mm/year.

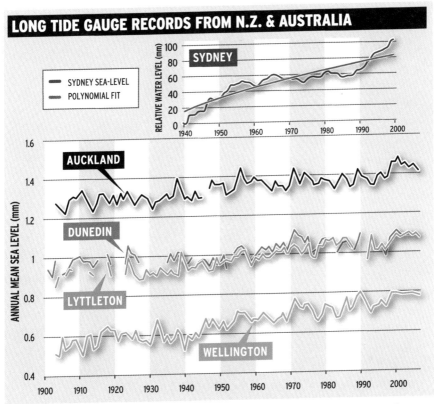

Fig. 25. 100 year-long tide gauge records from four NZ ports (Auckland, Dunedin, Lyttleton and Wellington), showing a progressive, irregularly varying sea-level rise at an average background linear rate of 1.8 mm/year (Hannah & Bell, 2012). Inset: 1940-2000 sea-level record from Port Denison (Sydney), showing similarity to the NZ records and (fitted polynomial grey line) that the average rate of rise of 1.1 mm/year over the last 60 years has been decelerating (Watson, 2011). These Australasian rates of sea level change are similar to the world average rise as estimated from a global network of similar tide gauges.

substrate as well as by the notional global sea-level. Local sea-level change occurs at greatly different rates and directions at different coastal locations around the world, depending upon the direction and rate of movement of the substrate.

Locations that were formerly beneath the great northern hemisphere ice caps 20,000 years ago, having been depressed under the weight of the ice, started to rise again as the ice melted.[29] Such isostatic rise continues today, for example in Scandinavia at rates up to 9 mm/year. Accordingly, local relative sea-level in such areas is now falling through time despite the concurrent

[29.] The process is called isostasy, and is caused by slow adjustment flowage in a hot, semi-plastic layer at depths of about 70-250 km, just below Earth's rigid outer shell (lithosphere).

long-term slow rise in global sea-level that has been driven by the melting glacial icecaps.

Conversely, at locations far distant from polar ice caps, such as Australia, no such glacial rebound is occurring, which results in local sea-level change in many places being similar to the global average rate of rise. Therefore, at many but not all locations around the Australian coast, sea-level has risen over the last century at rates between about 1 and 2 mm/year (Fig. 24, p.134; and see next section).

Tide gauge measurements

Local relative sea-level is traditionally measured using tide gauges, some of which have records as far back as the 18th century. These measurements tell us about the change that is occurring in actual sea-level at particular coastal locations, which includes rises in some places and falls at others.

Local relative sea-level observations are therefore the basis for practical coastal management and planning. Sea-level records up to 100 years long and more are available for both New Zealand and Australia (Fig. 25, p.135), and document long term rates of local sea-level rise that vary between about 0.9 to 1.6 mm/year. Recent study of these records by NSW scientist Phil Watson has shown, importantly, that in eastern Australia the rate of rise has been progressively lessening since 1940 (Fig. 25, upper right, p.135), rather than accelerating as required by dangerous AGW theory.

Satellite measurements

Since the early 1990s, sea-level measurements have also been made by radar and laser ranging from orbiting satellites of the TOPEX-Poseidon series. Situated in polar geostationary orbit, these satellites are able to make repeat measurements of the exact distance to the sea surface at particular locations every 10 days, as the Earth rotates below the satellite.

Averaging the repeat measurements for each location removes the effects of tides and waves, and yields an estimate of the height of the sea surface with an accuracy of about 2 cm. However, this accuracy is not fully secure because of lack of knowledge of the benchmark reference frame for the shape of the Earth (termed the geoid [30]), together with the need for corrections to be made for orbital drift and decay and for the stitching of records from different successive satellites.

---

[30] The geoid is a mathematical 3D-model of the shape of the Earth which takes into account differences in the density of rock materials buried below the surface. By definition the geoid represents a surface of equal gravitational attraction with respect to the centre of the Earth. Therefore at sea it corresponds to the mean surface of the ocean, whereas across continental areas it is represented by a smooth but undulating theoretical surface that lies at heights up to 70 m.

It is therefore not surprising that the satellite measures yield an estimate of the rate of global sea-level that differs from the tide gauge record, indicating an about double rate of rise of over 3mm/year. NASA's Jet Propulsion Laboratory (JPL) has recently acknowledged the importance of solving this mismatch problem by announcing a $100 million mission to launch a new GRASP satellite to improve the measurement of the Terrestrial Reference Frame that is used to calculate satellite sea-level measurements. JPL acknowledges that our current lack of an accurate model of the Earth's reference frame has introduced spurious (and unknowable) errors into all satellite-borne sea-level, gravity and polar ice cap volume measurements.

Calculating global sea-level change

Global warming in itself is only a minor factor contributing to recent global sea-level rise. This is because any warming is largely confined to the upper few hundred metres of the ocean. This surface mixed layer is constantly stirred and transported by surface winds, and even a 10°C warming in it would generate a rise in global sea-level of less than 10cm. Below lies the deep cold water of the ocean interior, but interchange of water, and heat, between the shallow and deep oceans occurs on long timescales of hundreds of years.

The second way in which global warming impacts on sea-level is by the melting of land ice — including both mountain glaciers and the ice sheets of Greenland and Antarctica — and the release thereby of extra water into the ocean. During the last interglacial, about 120,000 years ago, global temperature was warmer than today and significantly greater amounts of the Greenland ice sheet melted than today. As a consequence, global sea-level was several metres above the modern level.

The changing global sea-level generated by both ocean warming and ice melting is included in the computer models of the climate system. Ocean expansion can be directly related to warming of the surface mixed layer, but the melting of land ice is a more complex calculation that requires precise specification of surface temperatures. This is because ice does not necessarily melt if the surface temperature rises above 0°C; a small error can therefore make the difference between no melting and no sea-level change or actual melting and sea-level rise.

The range of surface temperature projected by the different IPCC models is shown in Fig. 20 (p.120) and underscores the considerable uncertainty that exists in model projections of future temperature and therefore sea-level rise.

Conclusion

Statements made about sea-level change by the IPCC and by government planning and management authorities in Australia and New Zealand nearly always refer to global sea-level. Policy is therefore being planned on the basis of unvalidated computer model projections, bolstered by the (probably inaccurate) satellite estimates of sea-level change. The accurate local sea-level data that is available from tide gauge measurements is largely ignored.

Much unrecognised uncertainty is thereby included in current policy planning. First, because of the uncertainty of the global temperature projections that feed into sea-level modelling; and, second, because of the lack of certainty also of the relationship between global temperature change and polar land ice melting rates.

Local relative sea-level change is what counts for purposes of coastal planning, and this is highly variable worldwide depending upon the differing rates at which particular coasts are undergoing tectonic uplift or subsidence. No evidence exists of dangerously changing trends in the available Australian or New Zealand tide gauge data.

What controls the position of the shoreline at Bondi Beach?
Not just sea-level, but also tectonics, sediment supply and weather.

Local sea-level is obviously an important factor that helps to determine exactly where a shoreline is located. However, along non-cliffed shorelines (i.e., gravelly, sandy or muddy beaches that front an estuary or coastal plain) three other important factors are also operative.

In the answer to the previous question (Why is it important to distinguish between local and global sea-level change?), we have already briefly explained one of these factors, which is the nature of movement in the underlying geological substrate. Other things being equal, a sinking substrate will result in a local sea-level rise and landward incursion of the shoreline; alternatively, an uplifting substrate will result in a local sea-level fall and seaward migration of the shoreline.

But such substrate rises and falls, and also any changes in global sea-level, take place along a land-sea interface that has varied characteristics. Where shorelines are made up of easily moved and transported gravel, sand and mud, the dynamic forces of wind, waves and tides cause the constant redistribution of sediment with concomitant changes in shoreline position and morphology. Thus the overall control on the position of a shoreline is the amount and direction of physical energy operative within the coastal system at any particular time. For example, daily tidal and wave movements usually operate to move sediment along a beach in the direction of longshore drift; and occasional heavy storms will cast some sediment high behind the usual high tide mark, and at the same time remove other sediment from the lower part of a beach to an offshore location. There usually being more storms in winter, more frequent storms then may result in a more landward location for the high tide mark, compared with quieter summer months when sediment may build up across the beach by accretion and cause the high tide mark to return to a more seaward position.

We are almost, but not quite, done, for there is a final factor that operates strongly on the position of the shoreline, and that is variations in sediment supply. Over time, the provision of sediment can exercise a dramatic influence on the location of the shoreline. For example, the delta of the Red River, Vietnam, has noticeably expanded in people's current lifetime, and the port of Ostia Antica, the harbour for ancient Rome, is now located several kilometres inland from the coast.

Mobile beaches are fed with sediment, most often sand, by the movement of material along the coast from sediment sources such as the mouths of rivers. As many coastal and harbour agencies have found to their cost, if you interfere with this shoreline river of sediment by constructing a groyne, seawall or port across it, the beach will build seawards updrift of the obstruction, and severe beach erosion problems will ensue at downdrift locations.

Gathering these thoughts together, and as every coastal dweller instinctively knows, sedimentary shorelines are dynamic geographic features. Their average position may shift landwards or seawards by distances of metres to many tens of metres over periods of time between days and years, in response to each of variations in the amount of sediment supply, the occurrence of periods of calm punctuated by major storms and variations in the position of local mean sea-level.

Is it true that Australia is going to be swamped by rising sea-levels?

Not on a societal planning time scale (centuries).

The fear about rising sea-levels swamping coastal properties in Australia and New Zealand, or even swallowing whole Pacific atolls, has been generated by two factors. The first is the misidentification of what causes coastal flooding today, and the second is the use of rudimentary computer models that project unrealistic estimates of future temperature and sea-level rise.

Modern coastal flooding is driven by the occurrence of rare natural events, most notably high spring tides, heavy rainfall over the interior and large storm surges, each of which can individually add a transitory metre or so to local sea-level height, or even 2-3 metres if combined — a height which can then be doubled for the storm surge associated with a very large cyclone. Such events often leave a temporary imprint at the coast such as shore-parallel lines of driftwood, shells or sand. Commonsense mostly guided early European settlers to build their dwellings landward of any such signs; today, planning regulations achieve a similar but more rigorous end.

The reality is, therefore, that if you choose to dwell in a beachfront property you are accepting the long term risk inherent in the vagaries of nature; and the fact that no tides or storms in living memory have covered your floor is no guarantee whatever that tomorrow's one-in-a-thousand-year storm surge might not do just that. Of course, destruction of natural coastal defences, such as dune systems, in the course of development will generally increase the local risks associated with erosion and inundation.

How then does the risk of future sea-level rise stack up against the already present hazard of flooding for anyone whose property is located within, say, 5 m vertically of modern sea-level? The answer lies in historical experience, as manifest by Figures 24 and 25.

During the last 100 years, the majority of locations around the Australian coast have experienced a sea-level change of between -20 cm and +30 cm. This amount is too small to have effected any noticeable changes within mobile beaches and shorelines that are constantly subject to the daily, seasonal and storm effects of weather variation and sediment supply (above: What controls the position of the shoreline at Bondi Beach?). From time to time, beach erosion or river outlet clogging makes the media headlines. Mostly the cause is a storm event, or natural or human interference with the longshore flow of sediment: sea-level rise or fall that might have occurred over previous decades has never been identified as a significant contributor.

In essence, and even when combined with the flooding and erosion risks already inherent in coastal locations, the likely sea-level change around Australia over the next 100 years is too small to require a major planning response. However, if the time horizon considered is expanded to the geological scale of, say, 1,000 years hence, and if current sea-level trends continue unchanged, then allowance will need to be made for changes of between -2.0 metres and +3.0 metres in AD 3010. It is, perhaps, a little early yet to be spending money on that distant, still small and anyway hypothetical problem.

What part does the ocean play in controlling climate?
Oceans are the flywheels of the climate system

Earth's climate system represents the transference of excess solar energy from the tropics to the polar regions by the circulation of the atmosphere and oceans (I: How does the climate system work?). These two circulations differ significantly in their capacity to transport heat because of their very different physical characteristics.

The atmosphere is a gas whose density decreases with altitude. Moreover, the density of the atmosphere is about one-thousandth that of the oceans; a full column of atmosphere is equivalent in mass to that of only the top 10 metres of the roughly 4,000 metre deep oceans. As a consequence of its greater density, the resistance to movement of the oceans is much greater than for the atmosphere. Once they are established, ocean currents tend to persist (i.e. they have inertia[31]), in contrast to the rapidly changing atmospheric motions that we observe every day in passing atmospheric systems.

Water being denser than air, the ability of the oceans to absorb heat (thermal capacity) is also much greater than that of the air, the stored heat of the entire atmosphere being equivalent to that of only the top 3.2 metres depth of the ocean. Because of the ocean's high heat capacity it is able to exchange much heat with the atmosphere without itself changing temperature. As a consequence, the heat content and temperature of the lower atmosphere are greatly influenced by any changes in temperature of the ocean surface layers. Cold air passing over warmer oceans is quickly warmed; conversely, warm air passing over cold water quickly loses energy and cools.

[31] Inertia: the tendency of an object to resist any change in its motion.

Despite its relatively low density, as air blows across the ocean surface it exchanges kinetic energy[32] with the ocean below. Not only are drift currents established but the surface layer of the ocean is also constantly stirred, mixed and moved in wind drift currents to a depth of about 200 metres. Solar radiation penetrates the ocean surface, and the incoming heat is then circulated throughout and tends to warm this upper mixed layer. As the layer warms, evaporation occurs at its upper surface, which in turn transfers heat and latent energy[33] to the lower atmosphere.

Because it is therefore a large heat reservoir, changes in ocean circulation, including the rate of mixing of cold subsurface water, can cause small changes in surface temperature that rapidly result in changed heat exchange with the atmosphere. This is readily apparent during El Niño events, when there is less mixing of cold subsurface water into the equatorial mixed surface layer, and ocean surface temperature increases. During La Niña events, when there is more mixing of cold water the surface temperature decreases.

The wind drag that stirs the surface ocean layer also transfers energy from the wind to the ocean, setting up wind-drift currents such as those that are driven by the trade winds. These ocean currents transport heat from the warmer tropics towards the poles, and also in east-west directions. In addition, wind drag can also generate local upwelling, which draws colder, nutrient and CO_2-rich deep water into the surface mixed layer, as exemplified by the upwelling Humboldt Current and its associated rich faunal diversity offshore from the west coast of South America. Such upwelling waters can significantly modify the temperature and chemistry of shallow ocean layers. As the surface temperature changes, so too the magnitude of local heat and gas exchange between the ocean and atmosphere varies. In consequence, the changing energy exchange between the ocean and atmosphere varies the amount of energy available to drive the local atmospheric weather systems.

The response time of the atmosphere and ocean circulations to surface exchanges of energy and momentum are different because of their different inertias. Successive storms blowing over a region of ocean will only slowly increase the ocean drift current and ocean surface temperature pattern because of the much larger thermal and mass inertia of the oceans. As the ocean surface temperature pattern varies, so too does the pattern of energy exchange with the overlying atmosphere change, and as a result there may be

---

[32] Kinetic energy: the energy possessed by a body by virtue of its movement, i.e., the energy of motion.

[33] Latent energy: heat energy released or absorbed by a substance during a change of state.

changes in the amplitude or position of regional Rossby Waves[34], and in the intensity of regional weather systems.

The natural meridional overturning of the oceans (called the thermohaline circulation, and with a time constant of around 1,000 years) is a major regulator of the annual climate cycle. But it is the interactions between the oceans and atmosphere, especially over the tropics, that contributes most to intra- and inter-annual variability of the climate system. The so-called El Niño South-ern Oscillation (or ENSO) is one such manifestation of ocean-atmosphere interaction (VII: What is ENSO and how does it affect Australian climate?), and its contribution to weather variability is second only to the solar-driven annual seasonal cycle.

Is there such a thing as ocean acidification, and should we worry about it?
No, the oceans have always been alkaline and will remain so, despite any minor decrease in alkalinity caused by the absorbtion of extra carbon dioxide

Acidity and alkalinity characterise the nature of chemicals dissolved in water. At extremes both are important because, in their different ways, acid and alkaline solutions are corrosive. Acids are characterised by a surplus of hydrogen (H+) ions whereas alkalines are characterised by a surplus of hydroxyl (OH-) ions.[35]

Most readers will be aware of the power of household cleaning fluids, and the need to use an appropriate cleaning agent to mop-up after par-ticular kinds of spill. Available cleaners range from alkaline to acid in nature, and gain their cleansing power from their ionic electric charge. Alkaline cleaners like soap, ammonia, and bleach are rich in hydroxide ions (OH-), and acid cleaners like lemon juice are rich in hydrogen ions (H+).

Natural fluids also exhibit a range of alka-line to acid properties, with rain (and fresh water derived from it) being mildly acid, distilled water neutral and sea water quite strongly alkaline.

---

[34] Rossby Waves are large meanders in high altitude westerly winds. The amplitude and location of Rossby Waves vary, and have a major influence on weather. For example, the waves are often exhibited along the polar jet streams and thus alter the cold-air flows that drive winter storm belts.

[35] An ion is an atom or molecule in which the total number of orbital electrons is not equal to the total number of protons in the nucleus, giving the particle a net positive or negative electrical charge and making it highly chemically reactive. Ca^{2+} and O^{2-} are examples of ions, in both cases carrying a double electrical charge.

Scientists characterise these properties using what is called the pH scale (for pondus Hydrogenus), which measures the concentration of H+ ions in a fluid. The scale runs from 1 (strongly acid; gastric fluid) through 7 (neutral) to 14 (strongly alkaline; strong bleach), and on it rain water generally measures 5.1-5.3, distilled water 7, and sea water varies between 7.6 and 8.3.

The ocean is alkaline, and has been so for at least the last one billion years because of its content of dissolved carbon dioxide and other alkaline salts. Given that broad chemical equilibrium has to be maintained between the carbon dioxide in the atmosphere and that dissolved in the upper levels of the ocean, adding extra carbon dioxide to the atmosphere inevitably results in an increase also in dissolved carbon dioxide. This interchange, and equilibrium, is dependent primarily upon temperature (carbon dioxide being more soluble in cold water), and less on pressure and salinity.

The ocean is the largest natural reservoir of dissolved carbon dioxide, and contains 38,000 Gt of dissolved carbon dioxide compared with just 770 Gt in the atmosphere. The ocean is strongly buffered [36] by clay minerals and ocean floor rocks, and in the presence of calcium ions (Ca_2+) also has the capability to sequester carbon dioxide into seafloor sediments by either chemical or biochemical precipitation of calcium carbonate ($CaCO_3$, as aragonite or calcite). In consequence, the ocean possesses a very large capacity to release or absorb more carbon dioxide, and also to resist changes in its pH, as atmospheric chemistry changes.

All of which said, it nonetheless remains true that as atmospheric carbon dioxide increases, so will dissolved oceanic carbon dioxide increase commensurate with maintaining chemical equilibrium across the water/air interface. The key question to assess, therefore, is what is the likely magnitude of surface pH changes under the likely future scenario of enhanced atmospheric carbon dioxide.

The answer to this question is summarised by Fig. 26. Figure 26 (left) contains a plot of Sr measurements from fossils corals that represent a proxy pH measure for the China Sea, which ranged between 7.9 to 8.3 over the last 7,000 years. Figure 26 (right) contains data points that represent pH measurements for modern water samples collected between 60°N and 60°S in the Pacific Ocean, and which exhibit a similar range in magnitude, this time from 7.8 to 8.5. Plotted atop the points in the right-hand graph are three horizontal lines that represent the modelled equilibrium pH level that will obtain

[36.] A buffer is a solution that resists changing its pH when acid or alkali are added to it, or when it is diluted with water. The tendency to stability arises because buffer solutions contain both a weak acid and a weak alkali which do not neutralise each other, and are therefore able to absorb (neutralise) any H[+] or OH[-] ions added to the system.

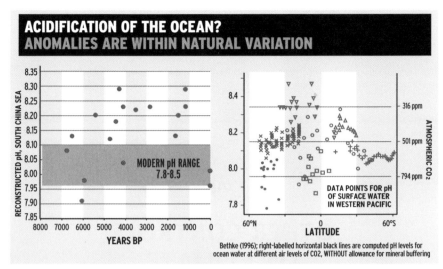

Fig. 26. Natural variation in ocean alkalinity (pH) indicated by (left) proxy-pH boron isotope data over the last 7,000 years from the South China Sea (modern pH range indicated by red shading bar) (Yiu et al., 2009); and (right) measurements in the modern Pacific Ocean between latitudes 60° N and 60° S; horizontal lines indicate the average pH level anticipated to occur at atmospheric carbon dioxide levels to 316, 501 and 794 ppm (Segalstad, 2013). Note that the natural fluctuations cover a range nearly four times the 0.1 unit decline in oceanic pH that is predicted to have occurred since pre-industrial times, and that any further changes will not result in ocean acidification but in an ocean that, although slightly less alkaline, will remain well within the range of natural pH variations.

assuming the single change of an increase in atmospheric carbon dioxide to 316, 501 and 794 ppm, respectively. The theoretical change in pH of 0.4 units for an almost 3-times increase in atmospheric carbon dioxide over pre-industrial levels can be compared with the typical pH swings of 0.3 units that occur naturally within shallow oceanic water on a monthly to yearly scale, and also falls within the envelope of natural variability of the modern Pacific Ocean. Furthermore, the calculations of future pH level take no account of the buffering presence of minerals in the ocean, and therefore very much represent a 'worst case' analysis.

In summary, the 'acidification of the ocean' hypothesis is based upon a kernel of scientific truth, but no possibility exists that the oceans will become acid. The reality is that any small increase in oceanic pH resulting from future increases in fossil fuel emissions will fall well within the range of natural variability of pH in the modern and past oceans.

Use of the term acidification implies to the general public that the ocean has switched, or will switch, to being acid. Given that the phenomenon actually being observed is instead a slight reduction in alkalinity, the use of such terminology is regrettable.

VII

OTHER CONTROLS ON CLIMATE

How important to global climate are sources of heat from inside the Earth?

Globally, not very; but measurable effects do occur locally near volcanic centres.

The dominant source of heat energy to the Earth is the Sun, whose direct radiation provides an average 340 watts/m² of heat at the top of the atmosphere (IV: Is the Earth in climatic equilibrium?). This incoming solar heat, or more strictly its redistribution by radiation and convection, is what drives the atmospheric and ocean circulations and hence regulates the global climate system (compare Fig. 3).

As anyone who has been down a mine knows, the deeper you go in the Earth the hotter it gets, and this fact must reflect the presence of a heat source at depth. This deep heat is derived from the molten core of our planet and by the decay of radioactive elements (for example, uranium, thorium, potassium) in the mantle and crust, and is transferred by slow conduction to the cooler surface.

Though Earth's internal heat-flow is readily measurable, its magnitude is so small that overall it has little effect on the global climate system. The range measured at the surface usually varies between about 20 and 40 thousandths of a watt (i.e., 20-40 milliwatts)/m², depending upon whether a measurement is made above oceanic (hotter, basaltic) or continental (cooler, granitic) crust. To put this into perspective, it represents roughly one-ten-thousandth the magnitude of solar heating.

These low average rates notwithstanding, geological processes sometimes cause the concentration of subterranean heat to the degree that rocks become locally molten, such as near active volcanic centres. When erupting, volcanoes can certainly have a significant effect on local weather patterns, causing cloud formation, lightning and rain.

More rarely, during particularly large eruptions such as that of Krakatau in 1883 and Mt. Pinatubo in 1991, fine particles that are thrown into the stratosphere may remain there to circle the Earth for more than a year. By reflecting incoming sunlight back into space, these fine-grained volcanic particles (called aerosols) can cause a temporary global cooling of more than 1°C at the Earth's surface, as indeed happened after the Pinatubo eruption in 1991.

So, in some rather oblique but nonetheless real ways, Earth's internal heat generation can indeed have an effect on climate. In global context, however, the climatic effects are either short-lived, minor or both.

How important are cosmic rays in affecting global climate?
Perhaps very important, but the influence of cosmic rays on climate has yet to be quantified.

In the 1990s, Henrik Svensmark, a physicist at the University of Copenhagen, proposed that the varying intensity of incoming cosmic rays[37], which is modulated by the constantly changing solar magnetic field, might be an important control on the formation of low clouds and thereby on Earth's temperature. The suggestion was, and remains, controversial, because

---

[37] Cosmic rays are high-energy particles that originate in galactic and deep space, or are emitted by the Sun, and impinge on Earth's atmosphere from the outside. Most incoming particles are protons (90%), the remainder being alpha particles (9%) and electrons (1%). The term 'ray' is confusing, as cosmic ray particles arrive individually, and not as part of a ray or beam of particles.

everyone 'knew' in the 1990s, and some scientists still believe now, that it is carbon dioxide that controls the temperature and climate.

Svensmark attempted to validate his theoretical ideas experimentally, using the sky as a source of cosmic rays and laboratory facilities available to him in Denmark. In what instantly became a classic experiment, in 2007 he showed that the high energy charged particles moving through air did indeed cause fragmentation of matter that they hit, producing clusters of particles about 3 nm in size. More comprehensive experiments have since been conducted using the powerful CERN particle accelerator, with the results again consistent with Svensmark's hypothesis.

So how might this be relevant to climate? The Earth is under constant bombardment from external cosmic radiation. As a cosmic ray penetrates the atmosphere, it often collides with an atom or molecule of atmospheric gas and causes it to fragment into a group of smaller particles. Further collisions and fragmentations occur in a particle shower that cascades downwards from the upper to the lower atmosphere. This is where low-level clouds form, by tiny water droplets condensing about particles of an appropriate size that are called cloud condensation nucleii (CCN). Typical CCN range in size from about 10-100 nm, which is significantly larger than the 3nm size of the particles produced by Svensmark's and the CERN experiments. However, in the real atmosphere, as opposed to a walled experimental chamber, it is possible that the clusters of smaller particles would grow within a few hours to the size of CCN.

What does this have to do with global temperature? Low level clouds are probably Earth's most important cooling thermostat, for their white tops reflect incoming solar radiation directly back to space. Therefore, cooling occurs as the area of cloud increases and warming when it decreases, with a worldwide change in cloudiness of just 1% causing a change of received energy at the Earth's surface of about 4 watt/m^2. This is roughly the same change that is caused by a doubling of carbon dioxide levels.

The next step in the chain of argument is an elegant one. The Earth lies within the magnetic field of the Sun, which from time to time waxes and wanes in strength. Characteristically, solar magnetic strengthening (called storms) is associated with sunspots and solar flares, and these magnetic pulses reach out and envelop the Earth. The final piece of the jigsaw is that the arrival of cosmic radiation in the upper atmosphere is strongly modulated by the combined strength of the Earth's own magnetic field and that of the Sun, incoming rays being deflected away by a strong magnetic field. So now, the denouemont — as the Sun undergoes a magnetic event, more cosmic rays

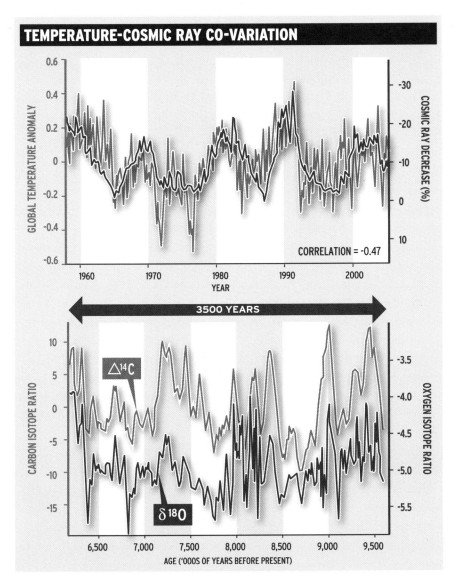

Fig. 27. Upper: Relationship between global temperature (blue curve; effects of volcanism, ENSO and a linear trend removed) and galactic cosmic ray flux (red curve) between 1960 and 2010 (Svensmark & Friis-Christensen, 2007). Periods of high cosmic ray flux (note inverted right-hand scale) correspond with periods of lower temperature and also (not shown) with lower solar activity. Svensmark's hypothesis is that this is because solar magnetic eruptions exercise a control over incoming cosmic rays, which themselves significantly contribute to the formation of low level clouds. Hence, lower solar activity, more cosmic rays, more clouds – and cooling.

Lower: Relationship between solar activity (blue curve; representing variations in radioactive ^{14}C) and rainfall (red curve; representing changes in oxygen isotope ratio) in 6,500-9,500 year-old samples from a cave stalagmite in Oman (after Neff et al., 2001). A close correlation is apparent between solar variation and climate in a manner consistent with Svensmark's hypothesis.

are deflected away from the Earth, so fewer CCN are available for the formation of low clouds, so the area of low cloud decreases, and so the temperature warms. Conversely, when the Sun is magnetically quiet, more cosmic rays penetrate the atmosphere, more clouds form and the Earth cools.

Strong empirical evidence also supports a link between cosmic radiation and Earth's climate. First, a strong relationship has been shown to exist between global temperature and cosmic ray flux between 1958 (when systematic measurements of cosmic rays commenced) and 2006 (Fig. 27, upper, p.149). Second, analysis of an early Holocene speleothem from Oman demonstrates a similar and very close parallelism between two proxy measures: one, the balance between oxygen isotopes, for rainfall; and one, the amount of radioactive carbon present, for solar activity (Fig. 27, lower).

In a final touch, Svensmark's hypothesis also offers an appealing explanation for the otherwise unexplained fact that as the Earth warmed during the late 20th century, and Arctic sea-ice retreated, so did Antarctica cool and its apron of sea-ice expand. The computer models that use carbon dioxide as the primary forcing agent for global temperature change predict that warming should occur concurrently at both poles. But if cloud rather than carbon dioxide is the controlling agent, then the inexplicable facts become predictable. For the dazzling whiteness of Antarctic ice means that it is the one large area on Earth where more low cloud causes warming rather than cooling, because the cloud tops have a slightly lower reflectivity than the ice.

The recent experiments and the strong evidence from geological data sets combine to provide credibility to Svensmark's ideas, despite the lack of experimental confirmation yet for every envisaged step in the process. IPCC scientists claim that solar variations are too small to provide an explanation for current climate change, but the cosmic ray hypothesis offers another real alternative to carbon dioxide as an important controlling agent for global temperature, and it therefore requires close and full evaluation.

What about the Sun?

The Sun is indisputably the primary energy provider for Earth's climate.

The Sun affects Earth's climate in several different ways. However, the IPCC is firmly of the opinion that although the Sun provides the primary energy input into Earth's climate, fluctuations in solar activity cannot have been the driver for the late 20th century warming. Many independent scientists disagree with this assessment.

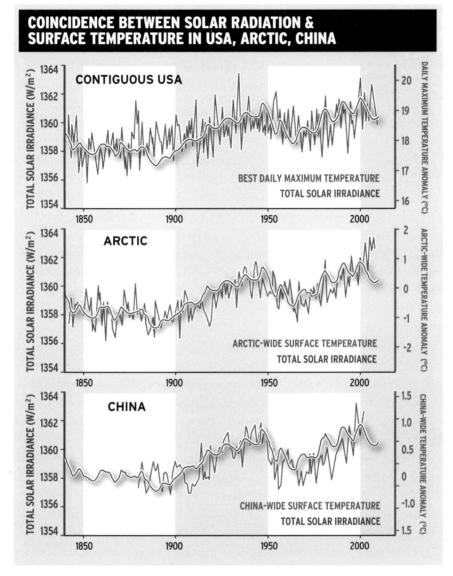

Fig. 28. A close relationship between total solar radiation and temperature exists for wide areas of the northern hemisphere (Soon & Legates, 2013), as manifest in USA (daily maximum surface temperature; top panel), the Arctic (surface temperature; middle panel) and China (surface temperature; bottom panel). These data suggest strongly that changes in solar radiation are helping to control temperature variations on at least a hemispheric scale.

In downplaying the importance of its influence, IPCC scientists point out that the variations in the visible radiation from the Sun (called the Total Solar Insolation, TSI) that occur in sympathy with the 11 year sunspot cycle are too small to have produced the 20th century warming on their own. So far as it goes, this argument is correct, but it ignores several other very important factors.

The first of these is that in addition to the 11-year cycle, the TSI of the Sun also varies on longer wavelength cycles, including importantly at periods of 80, 180 and about 1,500 years, which together control the occurrence of what are called Grand Solar Minima and Maxima. By happenstance, these longer cycles combined to produce a solar maximum towards the end of the 20th century that may well have influenced the warming seen then, despite the total increase in energy involved being only about 2 watts/m².

But a second, and even more important, point is that the Sun influences Earth's climate in several other ways than by variations in TSI. These other mechanisms include variations in the amount of solar energy provided in the ultraviolet and x-ray wave bands, and, as just discussed, through the modulating influence that its magnetic field exercises on incoming cosmic rays (above: How important are cosmic rays in affecting global climate?).

U.S. astrophysicist Willie Soon has made several vital breakthroughs in the understanding of solar-climate relationships. His recent research shows that for circum-Arctic locations as widely separated as USA, the Arctic and China, a strong and direct relationship exists between temperature and incoming solar radiation (Fig. 28, p.151), consistent with changes in solar radiation driving temperature variations on at least a hemispheric scale. Close correlations like these simply do not exist for temperature and changing atmospheric carbon dioxide concentration. In particular, there is no coincidence between the measured steady rise in global atmospheric carbon dioxide concentration and the often dramatic multi-decadal (and shorter) ups and downs of surface temperature that occur all around the world (compare Fig. 7, p.36).

Soon also advances evidence that the changes in solar activity control the volume of freshwater that is released into the Arctic Ocean, thereby modulating the 'conveyor belt' circulation of the great currents of the Atlantic Ocean and causing variations in the sea surface temperature of the tropical Atlantic after a 5–20 year delay. This time lag was not taken into account in earlier Sun/climate relationship studies, which probably explains their comparative lack of success.

Overall, the peer-reviewed scientific results about Sun/climate relationships are of disparate nature and suggest that there is a lot yet to be learned. The relationships outlined above have been obtained with independent datasets and stem from different and largely independent research groups. Considered together, the new solar relationships research suggests that it is premature to conclude that changes in solar activity play no (or only an insignificant) role in regulating climate.

What is ENSO and how does it affect Australian climate?
ENSO is climate rhythm that strongly influences rainfall in eastern Australia

ENSO is shorthand for El Niño-Southern Oscillation climatic variability. ENSO is centred in the Pacific Ocean, but has a worldwide influence and occurs at an irregular periodicity of a few years. The system oscillates between a La Niña state and an El Niño state. The relative strength of ENSO is measured using the Southern Oscillation Index (SOI), a meteorological term that is defined as the difference in atmospheric pressure between Tahiti and Darwin.

The importance of ENSO is its strong relationship with seasonal rainfall over many parts of the tropics, and also with temperature. The variation in global temperature between an El Niño phase and a La Niña phase can be as much as 1°C (compare Fig. 30, p.156), which is comparable to the long term change that occurred during the 20th century. ENSO is controlled primarily by the strength of the easterly trade winds that blow from South America towards northern Australia. For reasons that are not well understood, from time to time these winds strengthen abnormally and cause the system to enter the La Niña state.

During a La Niña episode (Fig. 29, upper), upwelling of cold water in the east maintains relatively cool ocean temperatures across the Pacific, and

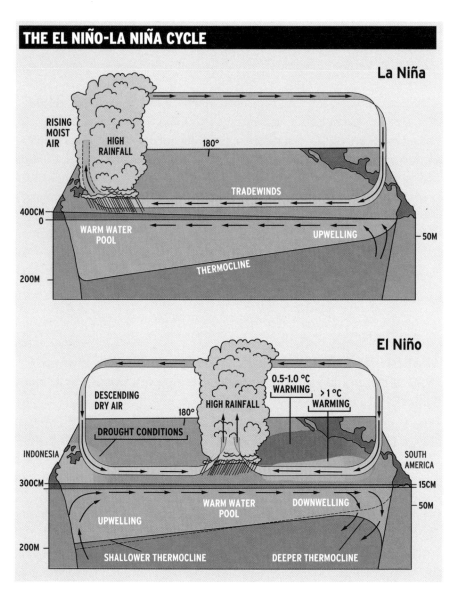

Fig. 29. The El Niño-La Niña climatic oscillation is a meteorological and oceanographic phenomenon centred in the tropical Pacific Ocean, but which has an effect on climate globally (figure after Skinner et al., 1999). For explanation, see accompanying text.

strong easterly trade winds drive warm surface water westward; consequently, strong evaporation and atmospheric convective activity occurs above the Western Pacific Warm Pool of water nearby to northeastern Australia. The warm pool can be as much as 5°C warmer and 65 cm greater in sea-level elevation than are the surface waters in the eastern Pacific. During a La Niña, deep tropical convection and heat and moisture exchange between the ocean and the atmosphere are much enhanced around Australia and Southeast Asia, causing flood rains. At the same time, the increased upwelling of cold water in the eastern equatorial Pacific results in aridity in coastal South America, and overall global temperature cools (Fig. 30, p.156). Once established, the anomalously cold waters in the central and eastern Pacific tend to persist for many months.

During an El Niño episode (Fig. 29, lower), upwelling lessens in the east and the equatorial trade winds slacken, thereby allowing the western Pacific warm water to extend eastward across the Pacific and reach the South American coast, and to spread north and south there; the focus of atmospheric convective activity moves also, to occupy a central ocean position. Though it only takes a few months for the warm waters to flow across the Pacific Ocean, taking the locus of deep atmospheric convection with them and causing a fall in sea-level in the western Pacific, it can take up to a year for the normal state of near-balance to be re-established thereafter. During an El Niño, drought occurs in eastern Australia and south-east Asia, heavy rain and local flooding occurs over the previously dry coastal regions of northern Peru and Ecuador, and an increase occurs in global temperature (Fig. 30, p.156).

Since 1980, significant El Niño episodes have occurred in Australia in 1982–83, 1987–88, 1991–92, 1993–94, 1994–95, 1997–98, 2002–03, 2006–07 and 2009–10; and La Niñas in 1988–89, 1998–01, 2007–08, 2008–09 and 2010–12. Most adult readers will know from their own experiences (and some more sharply than others) just how great the influence of this ENSO cycling is on eastern Australian weather patterns and climate-related natural disasters. The point is simply made by listing the following events.

- Flooding: NSW, 2000, 2007, 2011; Victoria, 2007, 2010–12; Queensland, 2000, 2008, 2010–11.
- Dust storm: Melbourne, 1983; eastern Australia, 2002, 2009.
- Drought: eastern Australia, 1979–83, 1995–2009; Queensland, 1991–95.
- Bushfires: NSW, 1994; Canberra, 2003; Victoria, 2003, 2006–07, 2009.

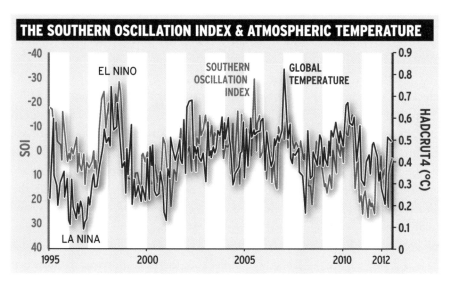

Fig. 30. Relationship between the Southern Oscillation Index (blue; a proxy for El Nino–La Nina activity) and global mean temperature anomalies (red) for 1995 to June 2012 (de Freitas & McLean, 2013). Because a lag occurs between a change in the oscillation index and the parallel change in global temperature, the plotted SOI has been shifted forward by 4 months. Periods dominated by El Niño conditions are associated with global warming, whereas periods dominated by La Niña conditions correlate with global cooling. The close correlation of the two curves implies that natural climate forcing associated with ENSO is closely linked with global temperature variability.

ENSO is rooted in the ocean–atmosphere interactions of the tropical Pacific Ocean, but its impact is global. For as the focus of convection shifts across the Pacific, the pattern of tropical overturning is disrupted, and the seasonal convection over equatorial Africa and the Amazon region of South America also changes in location and intensity. Moreover, because the focus of tropical atmospheric heating moves with the convection, ENSO impacts include a modulation of middle latitude weather patterns and a close relationship to global temperature (Fig. 30, p.156).

What is the Indian Ocean Dipole and how does it affect Australian climate?

Another climatic variability that significantly affects Australian rainfall.

The Indian Ocean Dipole (IOD) is another coupled ocean–atmosphere instability in which ocean surface temperature patterns affect Australian climate. The source is the equatorial Indian Ocean, and the return period is slightly longer than ENSO at about 5 years. The IOD is represented by an index that comprises the difference in sea surface temperature between the western and eastern equatorial Indian Ocean.

A positive IOD index reflects cooler than normal water in the eastern, and warmer than normal water in the western, Indian Ocean. A negative IOD index indicates the inverse situation. The effect of oscillations in the IOD is that a positive index is associated with decreased rainfall in parts of central

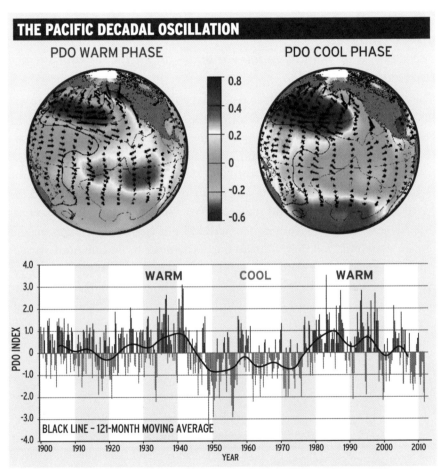

Fig. 31. Upper: Schematic maps of sea surface temperature (°C, colours) and wind stress anomalies (black arrows) typical of the two phases of the Pacific Decadal Oscillation (University of Washington). The Pacific Decadal Oscillation PDO is a 50-70 year long, ENSO-like pattern of Pacific Ocean climate variability that alternates between warmer than average North Pacific water temperatures (left figure; compare El Niño), and cooler waters (right figure; compare La Niña).

Lower: The PDO Index graph (monthly values, 1900-2012) is defined in terms of the North Pacific sea surface temperature variability poleward of 20°N (Giorgiog p2, 2013).

and southern Australia, whereas a negative index is associated with an increased rainfall over the same areas. Sometimes, but not always, negative IOD events occur in the same year as a La Niña event. Conversely, some but not all positive IOD events co-incide with El Niño years.

What is the Pacific Decadal Oscillation and how does it affect climate?

The PDO is a multi-decadal variation of Pacific Ocean surface temperature.

The dominant sources of short-term variability of Australia's climate are ENSO and IOD, with respective origins in the Pacific and Indian Oceans. However, all oceans have variability in their circulations that are reflected in changing surface temperature patterns. Such oscillations are most commonly multi-decadal, and three of the most studied are the Pacific Decadal Oscillation (PDO), the North Atlantic Oscillation (NAO) and the Arctic Oscillation (AO). The regional ocean surface temperature patterns that identify these oscillations are well established, though their origins remain uncertain and their climate impacts are yet to be fully understood.

Of these three oscillations, the PDO is perhaps the most relevant to Australian climate and also plays a global role. The PDO is linked to varying ocean surface temperatures in the North Pacific Ocean and has an apparent periodicity of 60-70 years.

The PDO is defined according to the variability present in the monthly sea-surface temperature in the North Pacific, and simulates a super-ENSO

cycle each phase of which lasts for 20–30 years (Fig. 31, upper, p.157). The PDO's cold phase (La Niña analogue) is characterised by cool waters in the central and eastern Pacific and warmer than usual waters in mid-high latitudes in the North and South Pacific Ocean. During the warm phase (El Niño analogue) this situation is reversed, with a pool of warmer than usual water occupying the central and eastern Pacific, flanked north and south by cooler waters. The index entered a new cold phase in 2008, and based upon PDO history since 1900, this suggests that the next 20 years will have cooler than average temperatures (Fig. 31, lower).

The major phases of the PDO are known to correlate with parallel changes in marine ecosystems, with the warm phase associated with enhanced biological productivity in the ocean near Alaska and lessened productivity, and a collapse of fisheries, off the west coast of the USA.

VIII

WHAT ABOUT
AUSTRALIAN CLIMATE?

Climate extremes: how hot does it get in South Australia and how cold in NSW?
Australia's recorded temperatures span 74°C.

Australia's hottest and coldest historic temperatures are +50.7°C, recorded at Oodnadatta, South Australia on January 2, 1960; and -23.0°C, recorded at Charlotte Pass, New South Wales, on June 9, 1994.

Over about the last 150 years, this represents a range of temperature across Australia of 73.7°C. These and similar local statistics represent climate extremes and their impact is as rare weather events. More typical daily temperature ranges are usually less than half of this, varying between as little as 5° in the maritime tropics in summer to perhaps as much as 30° in arid inland regions in winter.

It is a common claim that the number or magnitude of extreme hot or cold days, and contingent hazards, will increase because of the influence of human carbon dioxide emissions. For example, a 2006 report by CSIRO and the Bureau of Meteorology cautioned that the number of days in southeastern

Australia when the forest fire danger index will be very high or extreme is likely to increase by 4-25% by 2020 and 15-70% by 2050.

These, and similar claims, rest on projections by the rudimentary computer models of the climate system. Such claims must be treated as speculative, or scenarios, until the models are validated against independent data (V: But can computer models really predict future climate?).

In similar fashion, most living organisms are adapted to cope with regular daily and seasonal swings in temperature of 10°C or more, which belies other computer model projections that a modest 2° of warming will cause ecological catastrophe.

What controls Australia's rainfall?
A combination of geography, topography and seasonal meteorology

Rainfall at any location is regulated by two primary factors: first, the geographic position with respect to the major atmospheric circulations; and, second, the regional topography. The atmospheric circulations (Fig. 3, p.21) cause a strong cross-latitude (zonal) gradation of rainfall, from the copious rainfall of the equatorial trough regions, through the aridity of the subtropical high pressure zones and the wetness of the middle latitude westerly wind belts, to the low precipitation of polar regions.

Australia is a large island continent situated in the middle of a tectonic plate,[38] well distant from the earthquake and volcanic activity that marks plate boundary areas. This has not always been the case, and before about 140 million years ago Australia was joined to Antarctica as part of the southern super-continent of Gondwanaland. 35 million years ago, the creation of the Southern Ocean by sea-floor spreading between Australia and Antarctica caused Australia to finally break away from Gondwanaland and begin the slow northward-drifting trajectory that continues today at about 7cm/year. During all of this time the forces of erosion have been acting upon Australia's

---

[38.] The Earth's outer rigid lithosphere is divided into about a dozen large, 5-50 km-thick tectonic plates fitted together in jigsaw fashion and all in horizontal (and sometimes vertical) motion with respect to one another. Great earthquakes and explosive volcanism characterise the boundary zones where two adjacent plates abut.

older, interior mountain ranges such as Uluru and the MacDonell Ranges. It is therefore unsurprising that Australia's topography today is subdued when compared with that of many other large landmasses — and conspicuously so when compared with the nearby mountainous plate boundary areas of Indonesia, Papua New Guinea and New Zealand.

The latitudinal sweep of Australia is large, extending from the near equatorial tropics, through the relatively arid sub-tropical high pressure belt, and into the temperate westerly wind belt. It is therefore not surprising that rainfall patterns vary widely across the continent. Notwithstanding this variability, the prevailing influence of the precipitation-poor subtropical high pressure belt is paramount and the cause of general aridity. It is only during the summer months that monsoonal rains penetrate and bring abundant rain to the northern inland, and during the winter months that the storms of the westerly wind belt extend northwards to sweep across and bring regular rain to the southern parts of the continent.

Mountains act as important generators of rainfall because of the uplift that they impart to winds. As moisture-laden air is swept from the ocean across a landmass, the presence of mountains forces the air to rise and cool, and for clouds to form and precipitation to occur. It is for precisely this reason that the parts of Australia with the highest and most reliable rainfall occur along the geologically young Great Dividing Range that rims our eastern seabord. West of the range, the airflow is downslope and drying.

The relatively arid nature of much of Australia results, then, from the particular geography and geology of the continent, which is ancient, deeply eroded, generally low lying and distant from the ocean. From these facts follows another, which is that aridity and drought have been common features of the Australian continent throughout its existence. Throughout the last 20 or so million years, however, aridity has resulted from Australia's geographical location whereas the occurrence of droughts is an outcome of the irregularity of seasonal rain systems. The Australian summer monsoon is relatively weak and inconsistent when compared with those of Africa, South America and Asia; and if the monsoon does not develop, then the rains stay away. Southern Australia being on the northern margins of the westerly wind belt, it is also the case that any variation in the latter's seasonal movement will markedly impact on local rainfall. Drought occurs when seasonal rains are deficient, either because the monsoon has been weak or because winter storms have been absent.

Has Australia recently had more droughts than usual?
No.

Droughts are controlled by simple physical principles. During a drought, the land surface receives less rainfall than normal, and this causes drying of the soil. With less soil moisture available for evapo-transpiration, when the Sun's energy hits the surface less energy is partitioned into evaporative cooling and the surface heats more. Drying soils therefore feed into a warming feedback loop that produces higher surface temperature. Consequently, air temperatures during the day become more elevated than they would otherwise have been; and, because less moisture is returned to the atmosphere through evaporation, at the same time the humidity of the atmosphere drops, further accentuating the drying of the soil.

Thus the common statement that global warming will cause more droughts is the opposite of physical reality; instead, it is the drought and dry soils that cause higher temperatures.

1896 saw the start of the federation drought—extended extreme heatwaves
are not unprecedented. January 2013

More generally, evaporation is a powerful natural constraint to surface temperature increase, because evaporation causes cooling. The latent energy transfer to the atmosphere caused by evaporation occurs according to a relationship in which the heat transfer increases at an extremely rapid rate (near exponentially) as temperature increases.[39] In a classic paper published in 1966, Australian scientist Bill Priestley pointed out that evaporation is a critical constraining factor for local temperature. Peak ocean surface temperatures are about 30°C, in equatorial forests temperatures are constrained to about 35°C, whereas dry deserts can achieve temperatures greater than 50°C.

Modern observations indicate that a natural upper limit to the surface temperature of tropical oceans is about 30°C, as is indicated also by proxy records for ancient oceans. The near constant heat source of the Sun and the physical laws that regulate heat loss from the ocean surface determine this upper limit. Evaporation effectively acts as the Earth's cooling thermostat.

Finally, the most recent protracted drought in Australia, in 2001–2007, prompted much speculative media commentary about carbon dioxide-driven warming having caused or exacerbated the event. Rather than being exceptional, however, the recent drought was a relatively commonplace event, and similar to historic droughts that occurred in the later part of the 19th century (inspirational to writers such as Banjo Patterson and Henry Lawson), at the turn of the 20th century (Dorothea Mackellar) and in the 1930s. However, the climate record shows that much longer periods of drought than these have occurred in Australia in the past.

For example, studies of a Queensland coral core demonstrate the occurrence of a drought in the large Burdekin catchment that lasted almost 70 years, from 1801 to 1869 (Fig. 32, p.165). The black bar graph on the bottom right of the figure represents measured flow in the Burdekin River for 1920–2000. The central green line represents the amount of barium in a coral core that extends from 2000 back to 1761; in this context, barium serves as a proxy for land-derived river flood plume sediment. There is a strong correlation between high flow rates and high barium values during the period of overlap of these two records, which indicates that periods of drought in the Burdekin catchment are marked by the absence of peaks in the barium record. Based on these records and criteria, Queensland was in drought when Captain Cook sailed past in 1770, not long before that start of another more prolonged (69-year long) drought between 1801 and 1870. The generally dry early to middle

[39]. Called the Clausius-Clapeyron equation, and expressed as $dP/dT = L/T\Delta v$, where P is pressure, T is temperature, L is latent heat and Δv is the volume change of the phase transition from water to water vapour.

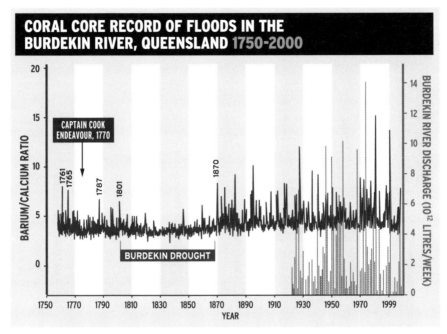

Fig. 32. Extended 19th century drought in Queensland, as revealed by barium measurements in a coral core (McCulloch et al., 2003). For explanation, see accompanying text.

19th century that is exemplified in this core contrasts with the warmer and wetter late 19th and 20th century.

Finally, in late 2012, Hamish McGowan from the University of Queensland and colleagues published a paper in Geophysical Research Letters which provided evidence that a major change in Australian aboriginal history may have been forced by a mid-Holocene drought that lasted for 1,500 years.

It is crystal clear that both droughts and lengthy droughts are part of Australia's normal climatic history.

Has Australia recently had more 'flooding rains' than usual?
No.

Assertions are often made that under the influence of global warming Australia has recently had more or more disastrous droughts and floods. Moreover, it is asserted that because of dangerous AGW such droughts and floods will become more frequent and extreme in the future. Such statements are generally accompanied by selected sections of the full climate record that fortuitously support the claim.

For example, graphs are commonly plotted that show a decline in average

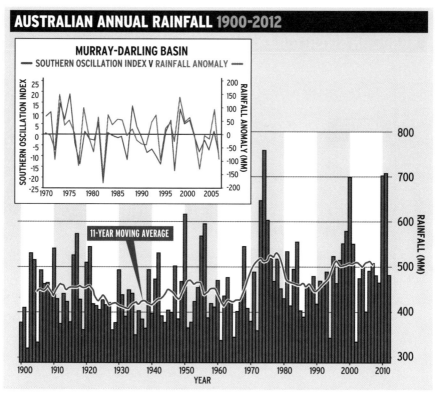

Fig. 33. Australian rainfall pattern, 1900-2012 (Australian Bureau of Meteorology). Note the lack of any increase in dry conditions across the record, or any pattern that would match with global temperature change or post-WWII increases in atmospheric carbon dioxide. Instead, both the annual data and the 11-year moving average indicate that wetter conditions than usual have prevailed since the 1970s. Inset: Plot showing the close relationship that exists between winter-spring rainfall in the Murray-Darling Basin and the Southern Oscillation Index, with heavy rains falling in eastern Australia during SOI-positive (La Niña) events.

rainfall during the second half of the 20th century. These graphs are accurate as far as they go, which is not far enough. Taken over the full record of the 20th and early 21st centuries, Australian rainfall has probably increased, especially over the tropical north (Fig. 33). Even the declining rainfall over southwest Australia during recent decades may not be unprecedented given that isolated 19th century records suggest that dry decades also occurred then.

Climatological cycles operate on many scales, out to geological timescales, but a dominant rhythm is that of the multi-decadal oscillation or variation that is apparent in long instrumental records (VII: What is the Pacific Decadal Oscillation?). Considering the full rainfall record for Australia (Fig. 33), it is apparent, first, that the decline in precipitation that occurred in the second half of the 20th century reflects the fact that the 1950s was a particularly wet decade. And, second, that the same wet period serves to anchor also a mirror-image pattern of increasing rainfall over the first half of the century. When the full record is considered it is seen to be extremely variable, in line with Australia's episodic, flashy rainfall and flood character. But over the full century, no change in precipitation is apparent outside of this natural variability, and certainly no trend occurs that could be linked with any confidence with increasing greenhouse gas emissions over the last 50 years.

The reason for the intermittent cycles of floods and droughts on both multi-decadal and several-year periodicity is that Australian rainfall, at least in the east and in the large Murray-Darling catchment, varies in close correspondence with the PDO, IOD and ENSO events (Fig. 33, inset) (see VII). Between 1945 and 1975, a negative PDO with frequent La Niña events (Fig. 31, lower, p.157) resulted in repeated widespread flooding across Australia. From 1975 until 2001, a positive PDO coupled with more frequent El Niño events meant that the flood risk was suppressed. Most recently, a return to PDO negative conditions and strong La Niña events has seen a return to the catastrophic flooding last seen in the early 1970's.

There is therefore no evidence that the recent 2009–2010 flood events were in any way unusual, given the long prior history of flooding in Australia.

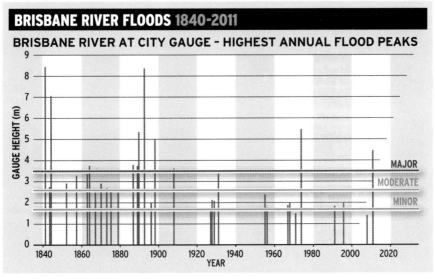

Fig. 34. Flood peaks in the lower Brisbane River, 1840-2011, as registered at the City Gauge (Bureau of Meteorology, 2011). Note the many large floods that occurred in the 19th century, and the lack of any trend of more frequent or larger floods since the mid-20th century, as has been predicted to result from increases in carbon dioxide emissions. The damaging 2011 flood ranks only 7th of scale of all floods since 1840.

For example, the 2009 flood of the Brisbane River, which at the time excited comment that it had been caused by global warming, ranks only 7th on the scale of all floods in the lower river since 1840 (Fig. 34, p.167).

Does the modern Murray-Darling system contain more or less water than at European settlement?
The basin now stores almost three times as much water as it did prior to settlement.

The issue of the water resources of the Murray-Darling Basin is a scientifically complex and politically vexatious one. We therefore do not presume to attempt a full commentary on the matter here. We do wish to note, however, a few fundamental scientific points that relate to climate and natural resource management, and that have been all but ignored in the present public debate despite their clear identification by biologist Jennifer Marohasy in her seminal paper, *Myth & the Murray*.

People's personal and political attitudes on the Murray-Darling issue are strongly influenced by their experience of the river as it is today, and has been

The environmental plan for the entire Murray/Darling river system depended on unproven computer model projections. October 2010.

Advocates of the dangerous AGW hypothesis appear to be tempted by the idea of media regulation. Green's leader Bob Brown and climate commisssioner Tim Flannery. March 2012.

in the recent past. But that experience is of a river that has been extensively modified to better manage the natural extremes of flood and drought, and to ensure regularity of water supplies. The modern river is therefore not a good natural frame of reference.

Famous historical photographs taken during several droughts show that prior to engineering works parts of the Murray-Darling river system sometimes dried up altogether, quite naturally. After one such drought, in the 1920s major engineering modifications were started along the river and proceeded with especial vigour between 1950 and 1980. A large number of dams and weirs were constructed in order to maintain river transport, to provide reliable water supplies for communities and to retain water for irrigation and other purposes (Fig. 35, p.170). Prior to the development of this infrastructure, riverine biodiversity was often impacted deleteriously by the extremes of natural flood and drought. The engineering works have not completely mitigated the impact of flood; there are still high river flows after heavy rain-

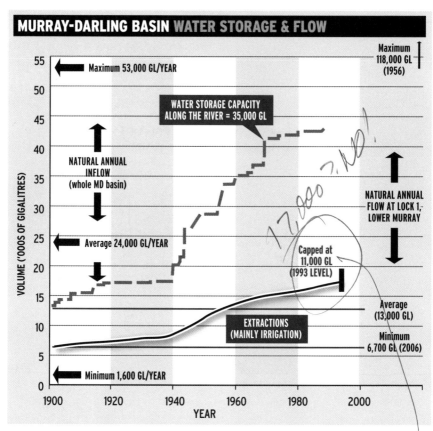

Fig. 35. Annual flow levels of the Murray River (red horizontal lines) superposed on a graph of water storage available in the river since 1920 (after Murray-Darling Basin Commission). Note that by the 1970s the river was able to store about three times the amount of water represented by the natural annual average flow. Note also that the dark blue curve representing water extractions has been arbitrarily moved up the left hand axis by 6,000 GL in order to improve the clarity of the plot.

fall events. However, the presence of the infrastructure has ensured a more regular river flow, which mitigates the worst impacts of both drought and flood in a generally environmentally beneficial way.

The result of these engineering modifications was that by 1980 the total volume of publicly managed water storage capacity in the Murray-Darling Basin was just under 35,000 GL. That amount is well over double the 13,000 GL that the river delivered as its average annual flow to the coast in its natural state.

By 1990, when extractions were capped, irrigation withdrawals from the Murray-Darling river system had reached about 11,000 GL, which comprises just 32% of the total storage capacity provided by water infrastructure. Effectively, the engineering works along the river have provided an economic source for irrigation water (at 1990 levels) AND provided for up to 24,000

GL of saved water. Importantly, the retained water has become the basis for more even river flow, and sustains both communities and riverine biodiversity during prolonged drought.

So long as the natural inflows to the river system do not diminish by more than 24,000 GL, the river system will remain better served than before the engineering works — even with current irrigation allocations retained. In reality, and despite the intermittent natural episodes of water feast and famine that have occurred in the basin since the 1890s, no evidence exists of any long term decline in basin rainfall or in the natural water inflow into the river system (Fig. 36).

The second way in which the Murray River has been significantly modified is by the construction in 1940 of the barrages upstream from the river's sea mouth, which transformed Lake Alexandrina into a fresh-water lake. Prior to the building of these barrages, the lower river in the vicinity of the present lake was an estuarine system within which saline water could penetrate into Lake Alexandrina seasonally (in autumn), and for longer periods during drought. During prolonged drought seawater from the Southern Ocean could and did penetrate as far as 250 km upstream from the coast.

Past natural flows through the Murray-Darling system were highly variable, ranging over the last 114 years from a high of 118,000 GL at Lock 1 in 1956 to only 7,000 GL in 2006. Even during drought, plentiful water always existed in the lower Murray River — being freshwater during floods, and

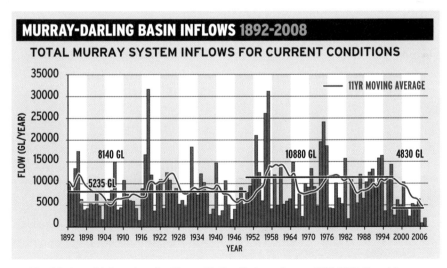

Fig. 36. Annual inflows into the Murray-Darling Basin catchment, 1892-2007 (data: Murray-Darling Basin Commission; CSIRO). Note the irregular distribution of times of high rainfall and drought, and the lack of any overall long term trend in inflow.

seawater and brackish estuarine water during prolonged drought. Erecting the barrages to create freshwater Lake Alexandrina has undoubtedly had the economic benefit of allowing farming to be developed around the lake. However, this has been at the expense of the pre-existing natural riverine environment, and has also produced subsequent strong political pressure to manage the upstream flows of the river in the way that since 1968 has supplied Lake Alexandrina with an average 6,000 GL of freshwater inflow a year.

As we said at the outset, the allocation of water rights and flows in the Murray-Darling Basin is a complex issue, but decisions about water use need to be firmly rooted in an understanding of the river system as it existed prior to human intervention. The extra 35,000 GL of water now able to be stored courtesy of beneficial engineering infrastructure on the middle and upper river systems provides irrigation water, a reliable water supply for communities and a buffer between extremes of flood and drought. In contrast, the construction of the barrages on the lower Murray River has been a less benign development so far as natural, near-coast riverine environments are concerned.

Has northern Australia suffered more, or more intense, cyclones lately?
No.

We covered in section III: What about other circumstantial evidence? the general claims that are made for an increase in the intensity or number of extreme weather events under the influence of dangerous AGW.

In Australia, one of the most dangerous of these weather systems is a tropical cyclone with its accompanying destructive winds, often torrential rain and, when making landfall, surging ocean level. Some have claimed that these storms have become more frequent and more intense over northern Australia as a consequence of dangerous AGW, and that their frequency and intensity will increase again in the future. Two types of evidence — historical observation and geological records — are relevant to assessing the question.

The Australian Bureau of Meteorology provides a list on its website of all cyclones that have made landfall in Australia since the late 19th century. Analysis of the list provides no compelling evidence for an increase in either the number or the intensity of cyclones in the second half of

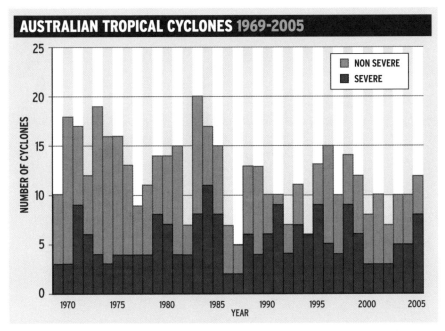

Fig. 37. Trends in tropical cyclone activity in Australia, 1969-2005 (Bureau of Meteorology). Note the declining trend in the overall number of cyclones since 1969. This is consistent with the global warming that occurred over the late 20th century, as such warming lessens the equator to pole temperature gradient that, among other things, spawns storms and cyclones.

the 20th century. Indeed, quite the opposite, for fewer cyclones have made landfall during the last 50 years than in the preceding time of record (Fig. 37). Writing in the *Journal of Geophysical Research* in 2008, Kuleshov and co-athors report that an analysis shows that, 'For the 1981–82 to 2005–06 tropical cyclone seasons, there are no apparent trends in the total numbers and cyclone days of tropical cyclones, nor in numbers and cyclone days of severe tropical cyclones.' These Australian patterns are consistent with the global pattern of tropical storms, which also shows a decrease rather than an increase in number of events since 1994 (compare, Fig. 14, p.86).

The second line of evidence bearing on the question comes from geological proxy data that record the frequency and magnitude of pre–European cyclones. Professor John Nott and colleagues addressed this issue in a 2007 paper in the journal *Earth & Planetary Science Letters*. Their research was based upon the analysis of oxygen isotopes from a speleothem in limestone caves near Chillagoe, north Queensland. The analysed record stretches from 1200 to 2000 AD, and, using the occurrence of heavy rainfall as a proxy for landfalling cyclones, it provides an estimate of both the number and intensity of

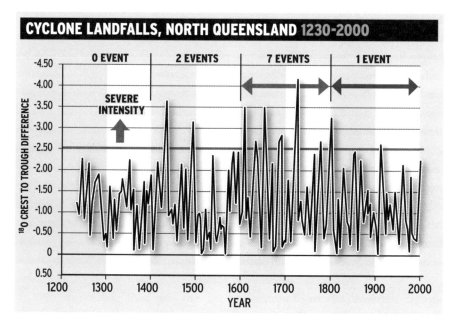

Fig. 38. Number of cyclones making land-fall in North Queensland that were of moderate to severe intensity, based upon oxygen isotope analysis of a speleothem from Chillagoe (Nott et al., 2007). Peaks above the grey horizontal line at -2.5 units represent cyclones of severe intensity. Note that more intense cyclones (7) occurred during the cold Little Ice Age, 1600-1800, than during the warmer period 1801-2000 (barely one severe event).These results are consistent with the data of Fig. 37, and with more and more intense cyclones occurring during cold periods.

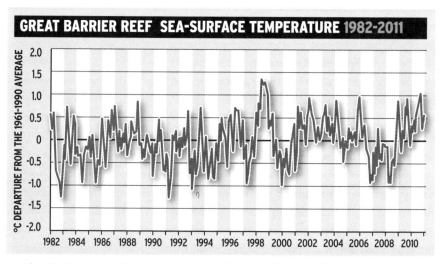

Fig. 39. Mean sea-surface temperature anomaly, Great Barrier Reef, 1982-2011 (NOAA, monthly values; McLean, 2011). Note the absence of any long term warming trend, and the general sympathy of the record with ENSO cycling, for example the conspicuous warming peak in 1998 (Fig. 9, p.75).

cyclones (Fig. 38). The Chillagoe record shows that no moderate to severe intensity cyclones occurred 1200–1400 AD, nine events occurred during the Little Ice Age period, 1400–1860, and only one cyclone of this intensity occurred during the warmer period 1860–2000. This evidence therefore indicates that over the last 800 years more and more intense cyclones have occurred in tropical Queenland during colder, not warmer, periods.

In summary, no evidence exists for either the occurrence of more or more intense tropical cyclones over recent decades of enhanced atmospheric carbon dioxide and warming temperature, nor for a future increase should significant global warming resume.

Is climate change destroying the Great Barrier Reef?
No.

The Great Barrier Reef (GBR) is an Australian environmental icon that is kept very much in the public eye by the wonders of its biodiversity and its tourism importance. The Australian government supports a major program for research and management of the reef system. As systematic monitoring of the reef became established during the 1980s, claims started to be made that the health of the reef was declining and possibly under threat.

Starting even earlier, in the 1960s, a seemingly endless list of threats to the survival of the reef has been advanced. First, crown-of-thorns starfish (COTS) infestations, allegedly caused by human influence, were reported to be destroying the living coral. Then sediment and nutrient runoff from the hinterland, allegedly a consequence of agricultural activity, was claimed to be threatening the reef with suffocation or pollution. More recently, bleaching of parts of the reef was observed and attributed to global warming. Finally, crown-of-thorns starfish infestations have again expanded in recent years, causing yet more comment about a dying reef.

In 1998 and 2002, severe coral bleaching occurred over large areas of the Great Barrier Reef. Fuelled by periods of atypically calm summer weather, water temperatures over shallow reef flats became unusually warm at the time, and the bleaching was blamed, quite wrongly, on global warming. Based on extreme IPCC projections, some ridiculous claims were made at the time that the reef system had only 10 or 20 years to go before it became a marine desert.

There is no doubt that a sudden increase in water temperature damages living coral, for bleaching events are observed to occur during episodic increases in local water temperature worldwide. However, bleaching is seldom a permanent outcome, because generally the coral recovers rapidly as temperatures return to previous values, as has proved to be the case for bleaching events on the Great Barrier Reef. Bleaching episodes are generally associated with EL Niño events, a combination of weaker winds and lower sea-level prevent cooler water from the surrounding deeper ocean from washing over the shallow coral reef flats. Notably, the regional sea surface temperatures in the reef tract area itself, though oscillating in sympathy with ENSO events, show no recent warming trend (Fig. 39, p.174). In reality, the recent bleaching outbreaks were not caused by global warming, but by localised surface water temperature increases resulting from hot and calm weather conditions.

Scientific studies published by Dr David Barnes and others in 2000 in the *Journal of Experimental Marine Biology* show that a statistically significant 4% increase in coral growth occurred on the GBR during the warming of the 20th century. More recent and detailed monitoring data collected by other staff of the Australian Institute of Marine Science (AIMS) in Townsville established that:

> data collected annually from fixed sites at 47 reefs across 1300 km of the GBR indicate that overall regional coral cover was stable (averaging 29% and ranging from 23% to 33% cover across years) with no net decline between 1995 and 2009.

Against this background of glowing reef health, a more recent report from AIMS, published as a US National Academy of Sciences paper and released on October 2, 2012, apparently contradicts the earlier research. The new paper claims that the Great Barrier Reef has lost half of its coral cover since 1985. Reading the fine print reveals that the researchers estimate that 48% of the coral loss that they report resulted from cyclone damage, 42% from crown-of-thorns infestations and the remaining 10% from coral bleaching events. Each of these three alleged causative agents is a natural part of the dynamic ecosystem that regulates the Great Barrier Reef. No connection is shown in the paper, nor in other current scientific literature, between any of these processes and global warming caused by carbon dioxide emissions, or, indeed, any other presumed human-imposed factors.

The reality is that late 20th century global warming exercised no discernible influence on the GBR. Though there is good cause to maintain a strong research and monitoring program in support of sensible reef management, the millions of satisfied tourists who continue to visit and enjoy its beauty every year attest that the health and biodiversity of the Great Barrier Reef remain unimpaired.

CODA

The seven questions addressed in this section are simply a subset of the more general question as to whether, worldwide, any increase in the magnitude, number or intensity of extreme weather events has occurred during late 20th century warming, and thus might be attributable to human-caused carbon dioxide emissions.

The first thing to say is that the framework thermodynamics of the climate system argue against the development of more, or more intense, weather events in a warming world. For, as long pointed out by Richard Lindzen, warming will result in an increased transfer of heat from low to high latitudes, and thereby cause a lessening of the pole-equator temperature gradient that provides the energy forcing for intense weather systems to evolve (compare, Fig. 3).

In any event, and moving from the theory to empirical evidence, in March 2012 the IPCC released a comprehensive report on this issue, which its scientists had spent several years compiling. Their conclusion was (our emphasis):

> There is medium evidence and high agreement that long-term trends in normalised losses **have not been attributed to natural or anthropogenic climate change** ... The statement about the absence of trends in impacts attributable to natural or anthropogenic climate change holds for tropical and extratropical storms and tornados ... The absence of an attributable climate change signal in losses also holds for flood losses.

The matter has been well summed up in a recent editorial in the general science magazine *Nature*, which concluded:

> Extreme weather and changing weather patterns — the obvious manifestations of global climate change — do not simply reflect easily identifiable changes in Earth's energy balance such as a rise in atmospheric temperature. They usually have complex causes, involving anomalies in atmospheric circulation, levels of soil moisture and the like.

Despite the many and continuing allegations made by reputed experts that contemporary severe weather events are a sign of human-influenced climate change, or are a harbinger of dangerous AGW, no empirical evidence exists in support of such views. Rather, the evidence indicates that intrinsic multi-decadal variability is the most important control over the pattern of severe weather events, which wax and wane in frequency and intensity as part of natural climatic variation.

The Wonthaggi water desalination plant in Victoria. During the European financial crisis governments faced the need to cut spending. December 2011.

IX

WHY DID WE NEED A CARBON DIOXIDE TAX?[40]

Why are economists involved in a scientific matter anyway?
Because they carry credibility in providing advice to governments.

Economics is the study of the distribution of scarce goods and services in an economy or, in a wider sense, the world. As a result, economists are often involved in resource allocation studies, which are aimed at determining the most efficient way to maximise the overall benefit to society. In such a context, economic studies contribute to every field of endeavour that requires the prioritisation of expenditure from within limited resources, such as taxpayer dollars.

Taxation has now been the major wealth distributive mechanism for more than ten thousand years. Taxation is used alike by ministers, politicians and dictators to pay for armed forces and social welfare, build palaces and employ hosts of public servants and tax gatherers.

[40.] This section has been written, and costs estimated, in terms of the reality of the present moment — which is that since July 1, 2012, Australian citizens and companies have been obligated by law to pay the costs involved in a carbon dioxide tax.

Perhaps surprisingly, economists often appear to carry more credibility than do scientists in providing advice to a government on a scientific matter, especially if it is a politicised one. In the context of current global warming politics in Australia, this is perhaps because economic advisers (being innocent of knowledge of the scientific issues involved) are judged to be less likely to challenge the official version of global warming science that is provided by the IPCC and favoured by politicians.

In any event, in October 2005, just as the global warming scare was getting up to full cry, British prime minister Tony Blair created the precedent of appointing distinguished economist Nicholas Stern to write an advisory report on the issue of global warming and related economic issues (released in 2007). Then Australian opposition leader, Kevin Rudd, duly followed the UK lead when he appointed leading economist Ross Garnaut in April 2007 to conduct a similar review for the Labor party, the final report being released in September 2008 after Rudd had become prime minister.

Predictably, both Stern and Garnaut took the (faulty) IPCC science on trust, producing compendious reports that contained carefully structured arguments

Carbon dioxide is plant food! June 2011

in favour of introducing carbon dioxide trading or taxation measures. Both reports are therefore of a political rather than objective or scientific nature.

The views expressed in the Stern and Garnaut reports have been strongly criticised for their naivety and inaccuracy by other professional economists and scientists. But this hasn't stopped governments and lobby groups from continuing to cite the reports, and the non-science-based views of other economists, in support of partisan lobbying for 'putting a price on carbon dioxide'.

Garnaut also introduced econometric modelling into the global warming debate. This is the economist's way of attempting to make economics 'scientific' by introducing static equilibrium models which exclude any difficult or contrary variables from the model. Such an approach completely negates an empirical, evidence-based analysis of the consequences of 'putting a price on carbon'. It ignores, too, the cascading consequences of a carbon tax, and the obvious gaming strategies and derivatives schemes that will arise as a consequence of trading in carbon permits, once it is allowed.

In short, econometrics is not a process of scientific inquiry. For example, it was just such an econometric model that was the major selling tool for the Collateral Debt Obligations that, in part, created the 2007 Global Financial Crisis. The model was written by very clever people, and looked impressive; but it was simply wrong.

Is the carbon dioxide tax a 'good' tax?
In a nutshell, no; and here's why.

Secretary of Treasury Ken Henry's mammoth 2010 Tax Review received almost 2,000 submissions from major companies and associations. The Labor Government of Kevin Rudd accepted just two of the minor recommendations in the review, despite the fact that Rudd was famously televised crooning over a pile of the Henry Reports saying, 'So much work'. Henry subsequently decided that it might be about time to leave the Public Service.

The Henry Committee followed global best practice in recognising that good taxes have most of the following five characteristics: equity, efficiency, simplicity, sustainability and consistency. However, the carbon dioxide tax possesses none of these attributes.

Equity
A consumption tax is intended to be equitable in the sense that only the user pays. But in cases where the tax covers an unavoidable consumption, as is indeed represented by a sizeable portion of household energy costs, the tax is regressive, i.e., unfair to people on relatively low incomes. Persisting with a

tax in such circumstances can only be justified on the grounds that it is targeted to discourage behaviour which the government deems to be undesirable, for example smoking.

In the case of a carbon dioxide tax, the inevitability of electricity and other consumption for routine daily living by poor persons makes it inequitable. In search of restoring equity, Labor's carbon dioxide tax is accompanied by significant offsetting payments to be made to low-paid people (defined as those earning between $6,000 and $18,000 a year) regardless of their individual, differing consumption patterns. Essentially, this and other concessions have been designed to offset the total carbon dioxide tax cost for those on low incomes. But for how long?

Estimating such costs is a fraught exercise. For example, an electricity increase of nearly 20% was recommended in May, 2012, by the NSW Independent Pricing and Regulatory Tribunal (IPART), which was twice the level of increase that the federal Treasury estimated would result from the new carbon dioxide tax. And even then, IPART's estimate doesn't include the cascade effect of the new tax as electricity users in turn pass on the increase to their retail customers.

Efficiency

Efficient taxes are easily understood and easy to collect.

If the carbon dioxide tax only lasts briefly, until it is either repealed by the Coalition or replaced by an Emissions Trading System by Labor, then it will be neither understandable nor easily collected. Even the value added tax, a simple tax that has had a long history in Europe, has resulted in Australia in thousands of pages of technical directives, and a very substantial body of court cases and rulings for individual goods and services.

The cascading nature of the costs of the carbon dioxide tax has caused software systems to be designed that ensure that the cost is always passed on to the next trading level. At this stage, the government has informed us mainly about consumer compensation for the tax, and about the bulk payments to be made to disadvantaged major manufacturers and power stations. We also know the parallel changes in personal income tax structure that are to be applied. But because there is no easy and accurate way to measure carbon dioxide emissions, the taxes are based on assumptions and models that can easily be in error by very large amounts in individual circumstances. Also, overseas 'carbon credits' are known to involve massive fraud and, in emissions terms, are therefore completely valueless.

At the end of 2012, the detailed mechanisms by which the Australian carbon dioxide tax is operating were still not understood by press commenta-

As a result of President Barack Obama's Australian visit it was agreed that a small US military force be stationed in northern Australia. November 2011.

tors or the public, despite the tax having been in place for six months. And in 2015, the tax is scheduled to be replaced by an Emissions Trading Scheme.

No tax can be regarded as efficient which is implemented without a clear functional and financial explanation, which is prone to fraud and which will be abolished in a few years (if Labor remains in government) in favour of an even more complex and uncertain carbon dioxide trading system.

Simplicity

The Labor Government has argued that the carbon dioxide tax is simple because only the largest 500 polluters were to be taxed. That number has now been successively reduced first to 320 and most recently to 294 business entities (which include bodies like large local Councils) that were listed on June 30, 2012 by the government's perversely named Clean Energy Regulator.

The cost of the tax obviously filters through the Australian economy, or at least it does for anyone who wants to use light and power. At every transaction level, from production to major users to transport to manufacturing to retailers and on to Australia's 24 million consumers, there will be unavoid-

With global warming or cooling there is always a business opportunity, June 2011.

able extra costs — unless, of course, you choose to sit in the dark, be cold in winter and hot in summer and go without food and drink. The extra costs are therefore not simply in the additional tax collected by the government, but also include this long line of additional on-costs.

To retrieve the headline tax, federal agents and the entire logistics chain of wholesalers and retailers in the Australian market must create appropriate accounting systems, and obtain advice from the Big Four Accountants (who are still recovering their Carbon Tax Professional Development costs for the past four years), or from smaller accountancy firms who have developed the tax technical expertise. All persons involved must create an audit and collection process that ensures that the tax is paid for the year or so until the tax is either terminated by a new Coalition government, or converted into an Emissions Trading Scheme by a re-elected Labor. If the latter path is followed, Australian carbon dioxide emitters and derivative speculators will be allowed to trade carbon dioxide permits internationally. There will be no compulsion to use the permits for their designated purpose, and they will be able to be bought and sold freely by both Australian and overseas speculators.

Sustainability

The current Minister for Climate Change, Greg Combet, has boasted that the carbon dioxide tax can't be abolished, but the tax will in fact be sustained only if Labor wins the 2013 election or the Coalition reneges on its promise to abolish the tax. These both seem unlikely scenarios.

We suspect that few citizens will have been impressed by government actions that aim to create another eternal tax to be added to the extant excise charges on alcohol and tobacco. Taxing alcohol and tobacco in perpetuity at least has the rationale that lives will be saved and health bills reduced. Taxing carbon dioxide, in contrast, is effectively taxing the very staff of life, given that the plants that it feeds lie at the base of most planetary food chains.

In Australia, a carbon dioxide tax will reduce everyone's living standard; a similar tax in third world countries, and the related ban on building coal-fired power stations, is likely to cause many deaths.

It seems clear that a carbon dioxide tax will prove to be an unsustainable policy development in Australia.

Consistency

Within Australia's so-called free trade market system, a whole menagerie of special industry assistance is now being directed towards the big companies who have been penalised by the introduction of the carbon dioxide tax.

This assistance includes hand-outs to offset price increases, changes in the tax threshold, special deals for some consumers, a programme of inquiries by the Productivity Commission that lasts until 2020 and, overshadowing all the rest, the spectre of an Emissions Trading Scheme in 2015.

A carbon dioxide tax is the very model of long term industry policy inconsistency. For 40 years, Australian governments have consistently held to the view that it is not their job to try to pick industry winners. Labor has now abandoned that position by creating a huge climate change industry that is intended to be directly funded by the Australian Government, using funds collected with a new and otherwise unjustifiable tax.

What is an Emissions Trading Scheme (also termed 'Cap and Trade')?

An ETS is an artificial, government-created market for carbon dioxide emissions trading.

An emissions trading scheme (ETS) involves the government creation of a market in which the supply of permits to emit carbon dioxide (called credits, and allocated by the government) is deliberately set by legislation to a level somewhat below the known demand, which is judged to be the total

As prime minister Kevin Rudd supported a cap-and-trade
system despite the inherent risks. August 2008.

pre-ETS sum of all industrial emissions. In Australia, this is to be achieved by allocating an original 500 (currently revised to 294) major emitters with credits that are equivalent to only a percentage, say 95%, of their continuing business-as-usual level of emissions. Who knows what deals have been cut for the other 206 businesses, or where they have gone?

Because carbon dioxide credits can be traded, the theory runs that those enterprises that can reduce their emissions cheaply to a level below their allocated credits will do so, with the intention of selling their spare credits to other enterprises that are unable to meet their own emissions targets. This, of course, creates a hedging market and an additional business risk for those companies that decide to hedge their carbon dioxide credit demand, only to find out later that they have woefully underestimated their true needs.

Basically, an ETS is an auction without any conditions or fundamental rules. People can game the auction. That means they can short sell, flood the market at a particular price and create scarcity for those who have a particular production need and who must have licences (at least up to the level of the penalty regime). Because the 'commodity' being traded is a colourless, odourless, tasteless and invisible gas, and is represented only by a piece of paper, corruption is inevitable.

Creating an artificial market for greenhouse gas emissions encapsulates the economist's dream of letting the agitation between supply and demand determine the price for permits. However, a pilot US trading scheme has already collapsed, and trials of the presently tottering but still just functional trading system in Europe have resulted in disastrously wide swings and roundabouts in the price of permits, accompanied by widespread corruption. A separate system operated by Russia and the Ukraine effectively acts as a second market for permit purchasers that always outbids the primary EU market.

Before any Australian system is integrated into either a European or a global system, which is the stated intention of Climate Minister Combet, a way will need to be found to prevent the enabling of a substantial additional market in carbon dioxide permit derivatives.

As this book went to press, on 16 April 2013 the European Parliament voted by 334 to 315 to reject modifications to the European emissions trading system that were designed to prevent its collapse. The immediate result was a 40% collapse in carbon dioxide market prices to below €3/tonne ($3.8/tonne). The survival of this market is therefore by no means assured, and it

July, 2009

may yet follow its US predecessor scheme (the Chicago Carbon Exchange) into oblivion. But if a European trading system does survive then the immediate implication for Australia is a budget income hole of many billions of dollars — engendered by the fact that legislation is already in place that from 2015 allows Australian businesses to buy their carbon credits on the European market at, say, $4/tonne compared with the $23/tonne currently mandated as the carbon dioxide tax price.

But surely we are just catching up with the rest of the world?

We are certainly told that, but it's simply not true.

A common justification that is given for the Labor-Green government's decision to proceed with carbon dioxide taxation has been that we are simply following the lead of other countries, and catching up with the world. In reality, the rest of the world's nations, including those of Europe, are still cautiously experimenting and thinking about the issue, with New Zealand, Japan, Canada and Russia recently withdrawing from the Kyoto Protocol, and the USA never having signed on to it in the first place. And, at $23/tonne, the current Australian carbon dioxide tax is being levied at nearly three times and ten times the 2012 rates in the European and New Zealand markets, respectively.

At government behest, substantial recent summaries of the way in which different countries are approaching 'putting a price on carbon dioxide' have been produced by the Productivity Commission and the Climate Commission. The more recent of these, from the Climate Commission, asserted that: 'Ninety countries representing 90 per cent of the global economy have committed to limit their greenhouse gas emissions and have programs in place to achieve this.'

Though this statement may be notionally true, none of the schemes listed by the Commission represent comprehensive, national carbon-dioxide pricing arrangements. Instead, most of the programs comprise incentive schemes for increasing the generation of renewable energy, or for boosting the energy efficiency of the construction or transport industries — significant aspects of which are no-regrets policies with their own intrinsic benefit, and have no necessary link to emissions reduction. Second, the specifically anti-emissions schemes listed by the Commission mostly represent an eclectic range of future promises, trials, industry and state pilot systems, and partial measures which penalise only particular users or industry sectors, exempt others (agriculture) and commonly provide protection or payment refunds for power intensive industries such as energy providers or metals smelters.

Meanwhile, at the collapsed Doha talks in December, 2012, only Australia and a handful of European states signed up for a continuation of a Kyoto Protocol type agreement. National governments have been unable to agree, under Kyoto or any other guise, to impose significant, as opposed to gestural pricing measures against carbon dioxide emissions. This reflects their unstated understanding that any such measures will have zero effect on future climate, and their common desire to alleviate the green political intimidation that they currently suffer as cheaply as possible.

Fig. 40 (p.190) summarises the Monty Python world that policy makers now inhabit in their belief that small cuts in OECD emissions will have a

Malcolm Turnbull loses the Liberal leadership as a result
of his support for carbon trading schemes. May 2011.

global effect. As the figure shows, whether western nations make further cuts in emissions is simply irrelevant to the rate at which global emissions will increase anyway, courtesy of the fast industrialisation that is occurring in Asia, South America and elsewhere.

It is therefore simply untrue to assert that the rest of the world is standing next to Australia on the issue. We are way out in front, and proportionally have already wasted a great deal more fruitless time and resources on the Climate Change Bet than has any other comparable nation.

Weren't MRET and other schemes already a de facto tax on carbon dioxide?

Yes, and to the tune of at least $15 billion/year.

The new carbon dioxide tax will yield income of about $9 billion/year.

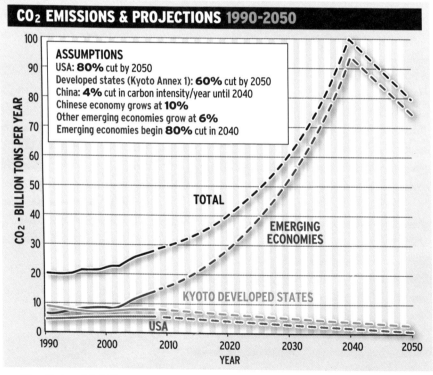

Fig. 40. Actual (solid lines, 1990-2007) and projected (dotted lines, 2008-2050) global totals of human-related carbon dioxide emissions (Muller, 2012). The projected amounts are probably an underestimate, because they were based upon the assumption that an extension to the Kyoto Protocol was to be agreed at the 2010 Copenhagen Climate Conference. Note that even if western industrialized nations were to have accepted further cuts, those cuts would have been pointless in the face of the fast rising emissions from the major developing countries.

Yet even before that tax was introduced, Australian business and consumers were already paying significant extra costs for various alternative energy and greenhouse initiatives, at both state and federal level.

The reason why so little has been said in the press about the large extra, and mostly hidden, costs to consumers for the provision of renewable energy is because the introduction of regulations such as the federal Minimum Renewable Energy Tariff (MRET) scheme, and related state initiatives, proceeded in bipartisan manner with support from both Labor and the Coalition.

Economist Alan Moran has estimated that, on top of the carbon dioxide tax, "the combinations of greenhouse emissions measures impose a cost in regulations and taxes of at least $15 billion a year and perhaps considerably more than this" – the measures referred to including MRET schemes, departmental budgetary expenditures, the $10 billion Clean Energy Fund, and various energy conservation regulatory measures .

This leads to what might be termed the Tony Abbott Conundrum, which is the decision that will be required when the carbon dioxide tax is repealed as to whether to junk the MRET system and associated special climate-related expenditures and regulations at the same time. Many persons argue that this needs to be done on the grounds that, like the tax, the MRET is simply an inefficient, ineffectual, environmentally damaging and swingeingly expensive gesture towards placating an imaginary carbon dioxide threat. However, needing to be done and being politically feasible are two very different things.

How much has it cost Australia to introduce the carbon dioxide tax?
An incalculable amount that will probably reach $100 billion by 2020

The demonisation of carbon dioxide emissions as a 'pollutant' goes back to the early 1990s. Since then, and in response to aggressive lobby group activity and strengthening public opinion, politicians of both major parties have been forced to grapple with the issue quite irrespective of their own personal opinions on the matter.

The government time (money) since spent on this issue at federal, state and local council levels has already been prodigious. In addition, manifold business interests, financial marketers and environmental lobbyists, amongst others, have spent a motza to promulgate their own interests and viewpoints.

The process has been irresistible, and no political or industry group has been able to afford the luxury of opting out, or 'waiting to see'.

Taxpayers, of course, have provided much of the very large amount of money that has been needed to sustain these preparatory activities and bribes. But these expenditures have just been preliminary, and since the implementation of the carbon dioxide tax on July 1, 2012, ordinary citizens themselves have been forced to absorb the flow-through costs of the new tax.

On top of the large preparatory costs, the direct cost of imposing a tax of $23/tonne from 2012 out to 2020 (given that Australia currently emits about 0.4 Mt of carbon dioxide annually) will be $83 billion. This estimate is a bare minimum, as increases of 2.5%/year in the price of carbon dioxide per tonne have already been announced, and emissions will grow, too, in line with economic growth. A final figure of $100 billion will therefore not be far off the mark. Of course, if the scheme is repealed, or the carbon dioxide price is refigured in line with lower international prices, then the direct cost will be concomitantly less.

There have been political costs as well as financial ones. Already, two prime ministers (John Howard, Kevin Rudd) and three leaders of the opposition (Brendan Nelson, Malcolm Turnbull, Tony Abbott) have either lost or gained their jobs partly or wholly over the carbon dioxide issue, and a third prime minister (Julia Gillard) may soon join them. No other political issue since Australian federation has had the potency to determine the fate of six successive federal party leaders. No wonder, then, that all politicians view the global warming debate as a highly toxic political issue.

Other costs of the brouhaha have included massive public service initiatives that involve thousands of federal, state and local body public servants, and a government policy-setting process that is much less stable than anything that has existed before. Finally, introduction of the tax has, in and of itself, undermined one of the historic advantages of the Australian economy (cheap, coal-fired power), thereby causing both a general increase in costs and also a substantial loss in international competitiveness.

What is the ongoing direct cost to me?
About $1,000/person/year, and increasing.

The carbon dioxide tax is not just a quick dip into the Salvation Army bucket that is passed around in the local pub. That makes you feel good, but is never enough to do the necessary job properly on a long term basis.

As Senator Barnaby Joyce likes to point out, every time from now on that you switch on an electrical appliance you will be paying an extra tax; every time that you purchase something in a shop, or from an Australian company over the web, you will pay extra tax; every time that you fill up the car with petrol, you will pay extra tax; and every time that you go on holiday, you will pay extra tax on all the expenses involved.

In effect, the carbon dioxide tax will act as a second consumption tax within the Australian economy, on top of the GST, and its inflationary effects will be felt on all transactions that are undertaken. The price increases will cascade right through the economy, and for most of them no compensation is proposed. At the bottom of the pile, to whom the accrued costs will be passed, lies the squashed citizen and consumer.

All of which said, there is no easy way of calculating the complete costs to the individual consumer of the carbon dioxide tax, nor of its possible ETS successor. It is therefore a bold economist who will attempt to predict what the net cost will be for the average consumer, but estimates have been made that the tax will add about 1.5% to the cost of living across the board. For a person on the 2011 Australian average weekly wage of $71,562 this equates to a cost increase of $1,073 per year.

Of course this cost will increase annually, in line with the scheduled 2.5% increase in the tax each year prior to the transition to an ETS. Beyond that generalisation, no one can predict with accuracy what the future costs will be.

Do the benefit of carbon dioxide trading schemes outweigh the risks of corruption?

No

Prior to the inception of the carbon dioxide tax on July 1, 2012, three historical examples of auctions of permits existed. They were the sale of imported motor vehicle quotas, the sale of textiles, clothing and footwear quotas and the auction for licences when Pay TV was introduced.

The government provided no household financial compensation for any of these schemes. In each case,

August, 2012

there was considerable exploitation of the auctions for large sums of money. More recently, for the carbon dioxide exchanges established overseas, corrupt trading has come to feature heavily.

As Bryan Leyland, one of the contributors to this book, has commented elsewhere:

> Carbon trading is the only commodity trading where it is impossible to establish with reasonable accuracy how much is being bought and sold, where the commodity that is traded is invisible and can perform no useful purpose for the purchaser, and where both parties benefit if the quantities traded have been exaggerated ... it is therefore an open invitation to fraud and that is exactly what is happening all over the world.

This view was supported recently by Australian Crime Commission executive David Lacey, who reported that the Italian mafia and other criminal groups already exploit the European trading scheme at huge profit, and that Australia now faces the same threat.

X

HOW WILL A CARBON DIOXIDE TAX AFFECT CLIMATE?

Will the tax fix the carbon dioxide 'pollution' problem?
What carbon dioxide pollution problem?

Although a tax may modify consumer behaviour by raising costs, it rarely fixes anything. Instead, it acts as a source of revenue for government. Also to the point, and as explained above (IV: Is Atmospheric Carbon Dioxide a Pollutant?), carbon dioxide is not a pollutant but an environmental benefice. There is therefore no carbon dioxide pollution problem that requires fixing.

For the purposes of the argument, however, let us concede that some persons do still believe that reducing levels of human carbon dioxide emissions would be a good thing to attempt. This is just the sort of problem that can be illuminated by undertaking a cost/benefit analysis of the competing policies that could be used to produce the wanted outcome. So surely we can just extract appropriate information from the numerous analyses of this type that must have been undertaken around the world, as governments have grappled with the global warming issue?

Well, actually, no. Remarkably, there is not even a single one.

Why not, you ask? Well the answer is simple, and it is that no effective analysis can be undertaken of a matter for which the measurement uncertainties are larger than the quantity that one wants to manage. Consider the following IPCC estimates for carbon dioxide emissions in 2005, expressed in billions of tonnes (Gt) of carbon (C)/year[41]:

Respiration (humans, animals, phytoplankton)	43.5-52
Ocean Outgassing (tropical areas)	90-100
Soil Bacteria, decomposition	50-60
Volcanoes, soil degassing	0.5-2
Forest cutting, forest fires	0.6-2.6
Anthropogenic emissions	7.2-7.5
Total	192-224 Gt C/year
Uncertainty	32 (~15%)

Canadian climatologist Professor Tim Ball was the first to point out that this range of estimates of natural and human carbon production (as carbon dioxide) has an uncertainty factor of 32 Gt C/year. The human contribution of 7.5 Gt lies within the uncertainty range of each of the first three natural sources, and the total uncertainty is almost five times the human production.

To put these numbers even further into context, consider that the Australian government's policy of taxing carbon dioxide to try to reduce emissions by 5% by 2020 is projected to result in a cut of only an insignificant 43 Mt of C/year. It is simply not possible to analyse, nor to manage in any rational way, components of the global carbon cycle that are so small that they are dwarfed by the uncertainty margins of the largest natural sources and sinks.

What percentage of carbon dioxide does Australian society generate?
Depending upon how you phrase the question, somewhere between 0.0001% and 2.7%

In a now famous programme made on March 15, 2011, 2GB broadcaster Alan Jones commented that Australians produced just '1% of .001 per cent of carbon dioxide up there', which is equivalent to 0.00001%, i.e., one part in ten million.

The statement caused a storm of media criticism, with most reporters promulgating an alternative estimate of Professor David Karoly's that 0.45% of all atmospheric carbon dioxide is in fact sourced from Australia. Somewhat surprisingly, Jones' statement was referred to the Australian Communications

---

[41.] Confusingly, discussions of cutting carbon dioxide emissions, or of fluxes related to such cuts, are often conducted in terms of millions (Mt) or billions (Gt) of tonnes of carbon (C)/yr. Accounting for the different molecular weights of C and CO_2, the multipliers for converting amounts of C to CO_2, or amounts of CO_2 to C, are 3.7 and 0.27, respectively.

and Media Authority (ACMA) for a ruling as to its correctness. In a decision in October 2012, ACMA ruled that Karoly's answer was the right one, and directed that Mr Jones undergo remedial training in factual presentation and that his program employ a fact-checker for material before it is put to air.

So why such a large difference between these initial estimates, and also a later Karoly estimate of 0.00018% that he made in an interview and letter interchange with Jones? Which one, if any, is correct?

Australian emissions		as a percentage of				= answer
A	B	C	D	E	F	G
Australian annual emissions (as % of D)	Australian cumulative emissions (as % of F)	Natural annual emissions (as % of all emissions)	Human annual emissions (as % of all emissions)	Human cumulative emissions (as % of atmos)	Total CO_2 in atmos (as % of atmosphere)	*Australian proportion of emissions (calculated in % rounded)*
0.4 Gt 0.05 ppm	13.1 Gt 1.43 ppm	743 Gt 95 ppm	29 Gt 3.6 ppm	859 Gt 110 ppm	3046 Gt 390 ppm	(Data from CDIAC, IPCC 4th AR)
1.3%		96%				0.013 x 96 = 1.2%
	0.43%	96%				*0.0043 x 96 = 0.4%*
1.3%			4%			0.013 x 4 = 0.5%
	0.43%		4%			*0.0043 x 4 = 0.2%*
1.3%				0.01%		0.013 x 0.01 = 0.0001%
	0.43%			0.01%		*0.0043 x 0.01 = 0.00004%*
1.3%					0.04%	0.013 x 0.04 = 0.0005%
	0.43%				0.04%	*0.0043 x 0.04 = 0.0002%*

Table 4.[42] Table showing varied ways in which Australia's contribution to global carbon dioxide emissions can be calculated, the eight alternatives each depending upon particular and different assumptions, i.e., representing answers to different questions. The three probable estimates of Jones and Karoly correspond to the entries in the yellow shaded cells (see text for fuller explanation).

[42] 1 ppm = 7.81 Gt of carbon dioxide; or, 1 Gt = 0.13 ppm of carbon dioxide.

The lack of correspondence in the three answers is because the two authors were calculating three different things. That this was the case resulted from ambiguities in Jones' original statement, which can be taken to refer to either the annual Australian carbon dioxide contribution today, or to the cumulative magnitude of Australia's contributions since the start of industrial emissions. In addition, the phrase 'up there' adds further ambiguity, for it might refer to either the annual or the accrued human emissions in the atmosphere; or to the annual or the accrued natural emissions in the atmosphere; or to the carbon dioxide content of the whole atmosphere itself.

Given these various ambiguities, it is perhaps not surprising that a seemingly simple question about Australian carbon dioxide emissions can be interpreted and calculated in at least eight different ways, as shown in Table 4. First, Australian emissions can be calculated using either a single current-year figure (0.4 Gt of CO_2, column A) or an accrued-since-1751 figure (13.1 Gt of CO_2, column B). Second, each of these figures (expressed as a percentage of all human annual emissions or the total CO_2 in the atmosphere, respectively) can then be compared with the four different carbon dioxide totals listed in columns C to F, yielding answers that range between 0.00004% (4 parts in 10,000,000) and 1.2% (1 part in 100). All of these different answers are 'right' with respect to the (differing) questions that they address.

It immediately catches the eye that one of these answers (right side, 5th line in data Table 4) would correspond to Alan Jones' original statement if it is assumed that one decimal point was accidentally misplaced. Effectively, and assuming that he did indeed intend to refer to Australian annual emissions as a proportion of the cumulative total of all human-related emissions, Jones asserted that the Australian carbon dioxide contribution was one part in ten million instead of one part in one million. Such a slip is obviously unfortunate, but both numbers anyway being very small ones, it is scarcely the hanging offence that has been made out.

What then of the 0.45% and 0.00018% estimates that are listed in Karoly's email to Alan Jones?[43] The first of these corresponds to Column B in Table 4. It is a little higher than the 0.43% indicated in Column B because Karoly's starting figures are slightly different to ours.[44]

The second figure of 0.00018% was explained in an email that Karoly sent to Jones after an interview in May, 2011, and represents the bottom line in Table 4. The calculation is accomplished by taking Australia's estimated

[43.] http://www.abc.net.au/mediawatch/transcripts/1116_karoly.pdf.

[44.] Karoly's 0.45% figure represents Australia's annual emissions of ALL greenhouse gases, expressed in terms of carbon dioxide equivalent; our 0.43% estimate is based upon Australia's carbon dioxide emissions alone.

cumulative carbon dioxide equivalent of 0.45 Gt (Karoly's 1.5%) and multiplying it by the 0.04% number entered in column F, to give the Australian contribution as a fraction of the total atmosphere. Karoly's result from this calculation is 0.00018%, which is effectively the same as our calculation of 0.00017%, both estimates rounding to the 0.0002% that is entered in the bottom right cell of Table 4.

Whoever would have thought that such an apparently simple question as the one posed at the head of this section could have such a confusing variety of answers? Apparently not ACMA, whose judgement reveals no understanding of scientific uncertainty, or of the ambiguities implicit in the matter that they were considering. ACMA simply made an authoritarian ruling on Alan Jones' statement based upon its own unrevealed thinking, and on uncritical acceptance of Karoly's estimated figure of Australia's emissions percentage.

In the implementation of the carbon tax, the need to maintain power capacity ensured that the federal government must compensate coal fired power stations including the so-called 'dirty' Hazelwood. September 2012.

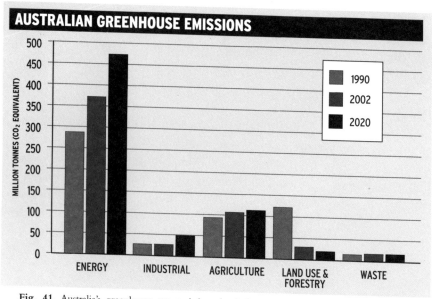

Fig. 41. Australia's greenhouse gas emissions by industry sector in 1990, 2002 and 2020 (projected) (National Greenhouse Gas Inventory, 2002).

How large are Australian carbon dioxide emissions in a global context?

Very small: and, anyway, over all of Australian territory we absorb as much as 20 times more carbon dioxide than we emit

It is not uncommon to read or hear in the media claims that Australia produces the largest emissions of carbon dioxide per head in the world. These claims, and variations on them, are false.

In 2008 the world's four largest emitters were China (7.0 Gt), USA (5.5 Gt), the European Union (4.2 Gt) and India (1.7 Gt), who together contributed 61% of total human-sourced carbon dioxide emissions. Australia ranks 16th on the list with 0.4 Gt/year, which is equivalent to 1.3% of global emissions, the breakdown of which is shown in Fig. 41.

Alternatively, if emissions are calculated on a per capita basis, then the four top ranking emitters are Qatar (53.5 t/person), Trinidad and Tobago (37.3 t/person), United Arab Emirates (34.6 t/person) and Netherlands Antilles (31.9 t/person), with Australia in 11th position at 18.9 t/person).

In any case, all such published statistics are calculated for the purposes of the Kyoto Protocol and related matters. They are therefore political in nature and largely divorced from scientific reality, because the Protocol deals only with land-based emissions and sinks.

From the scientific viewpoint, the contribution that any nation makes to the atmospheric balance of carbon dioxide must be assessed in the context of all natural sources and sinks. By leaving out the ocean, the Kyoto Protocol became effectively meaning-

less. This is because the ocean is by far the largest long-term sink of carbon dioxide on the planet, containing 144,000 Gt of dissolved gas, 3,700 Gt of which is located in the shallow ocean where it can readily be exchanged with the atmosphere.

Photosynthesising by phytoplankton in the surface ocean is a major mechanism whereby about 185 Gt/yr of carbon dioxide is transferred from the atmosphere to the ocean globally, which coincidentally is about equivalent to the 190 Gt/year sequestration that occurs in growing land plants. Marine plankton make this large annual contribution because of their short life time of about a week, which results in their replacement about 45 times every year; in contrast, land plants reproduce themselves only about once every ten years. Some of the carbon dioxide absorbed by phytoplankton returns to the atmosphere when the plankton die, but some also is lost to the deep ocean as sinking sediment particles. This process acts as a biological pump, whereby carbon dioxide is transferred from the atmosphere to the deep ocean. The pump is most active, and therefore effective, in the cold, high latitude waters that are preferred by important groups of phytoplankton such as coccolithophores.[45]

More broadly, the amount of carbon dioxide that can be chemically dissolved in the ocean depends upon the temperature, with more absorbtion occurring the colder the water and more outgassing occurring from warmer, tropical water.

Territory that Australia claims stewardship over, both land and marine, stretches from almost the equator to the South Pole and includes a huge area of carbon dioxide-absorbing Southern Ocean. Leaving aside the Antarctic claim of 2 million km² of extra ocean territory, Australia has acknowledged jurisdiction over 8.1 million km² of Exclusive Economic Zone

[45]. Coccolithophores are tiny floating marine plants that have a calcium carbonate skeleton up to 100 μ in size. Numberless plates of the skeleton sink to the deep floor on the death of the plants, where they form a marine ooze that lithifies into chalk when buried (the White Cliffs of Dover being a famous example of this; the Gingin Chalk of Western Australia and Amuri Limestone of Marlborough, New Zealand are other local examples).

(EEZ) ocean area, which is more than the continental landmass area of 7.7 million km^2.

Australia's current annual output of carbon dioxide is 400 Mt. The average rate of transfer of carbon dioxide from the atmosphere to the ocean by phytoplankton is about 512 tonnes/km^2/year. To a first approximation, therefore, Australia's 8.1 million km^2 area of EEZ will absorb these emissions 10 times over, and 12 times if the Antarctic EEZ is taken into consideration. Note that the ocean estimates are conservative, because much of Australia's EEZ comprises cold southern ocean which is richer in phytoplankton than the world average. On top of which, Australia's land plants probably sequester about the same amount of carbon dioxide again as the ocean phytoplankton do.

Thus in terms of balancing the planet's carbon dioxide budget, Australia is already doing its fair share of sequestering excess carbon dioxide emitted by landbound nations, and in doing so is punching much above its 22 million persons' weight. Exactly similar arguments apply to New Zealand, which has jurisdiction over 4.1 million km^2 of absorptive Southern Ocean against which to offset the emissions from a population of five million peole living on a landmass 15 times smaller in area than its EEZ.

How much warming will be averted by cutting Australian emissions?
An unmeasurably small amount.

Any cost/benefit analysis of the value of penal actions against human-related emissions of carbon dioxide must weigh up the costs of taking an action against the presumed benefit of the reduction in global temperature (more strictly, the amount of warming prevented) that will result from a specified cut in emissions. In Australia, both major political parties have adopted the same target, which is of a 5% cut in emissions by 2020.

Remarkably, neither party has stated publically what cooling will result from

a cut of this magnitude. Even more astonishing, most Australian journalists and media commentators have failed to pursue the matter despite repeated promptings by independent scientists. In principle, determining the magnitude of the warming prevented is a simple matter, so let's explain how it is done.

Because carbon dioxide is a greenhouse gas, increasing its content in the atmosphere will cause temperature to increase according to a simple mathematical relationship established by the IPCC.[46] Importantly, the relationship is logarithmic (IV: Is less warming bang really generated for every extra carbon dioxide buck?), which means that each successive incremental increase in carbon dioxide produces a diminishing rather than a fixed increase in temperature. Secondly, the x-term in the relationship that determines how much warming will occur is called the climate sensitivity, and corresponds to the increase in temperature that will result from a doubling of carbon dioxide (IV: What is climate sensitivity?). The magnitude of climate sensitivity is not known with certainty, but is assigned a value of 3.3° C by the IPCC in their Fourth Assessment Report (2007).

Using the IPCC equation and assumed climate sensitivity, carbon modeller and former Australian Greenhouse Office staff member, David Evans, has estimated that the following reductions in warming would result from differing cuts in Australian emissions up to 2050:

Reduction in Australia's emissions, 2011 to 2050	Resulting decrease in global temperature in 2050
0%	0.0000°C
5%	0.0007°C
10%	0.0015°C
20%	0.0031°C
50%	0.0077°C
100%	0.0154°C

The two most important points to note are, first, that meeting the targeted 5% cut by 2020 would result in a miniscule and unmeasurable 7/10,000°C of warming averted. And, second, that if we shut the entire Australian economy down so that we emitted no carbon dioxide starting tomorrow, the temperature in 2050 would be just 15/1,000° C cooler than continuing with business as usual. The projections underscore the futility of Australia acting alone in the absence of a global agreement.

These astonishing figures result from calculations using the IPCC's value of 3.3°C for climate sensitivity, a figure that nearly all independent scientists think is significantly too high. Making similar calculations to those of David Evans, but using a more reasonable value for sensitivity, Lord Christopher Monckton has estimated that the same 5% cut in Australian emissions

---

[46.] The increase in temperature (ΔT) for an increase in atmospheric carbon dioxide level from C_0 to C is: $\Delta T = x \star \log (C/C_0)$, where x is the climate sensitivity and is estimated by the IPCC (2007) to be 3.3°C for a doubling of carbon dioxide concentration. Other researchers, using empirical evidence, suggest a sensitivity of less than 1°C for doubling carbon dioxide (compare Fig. 17).

will result in 0.00007°C (about 1/14,000°C) of warming prevented. Or, put another way, the carbon dioxide level in the atmosphere in 2020 would be reduced to 411.987 ppm compared with the 412 ppm that it would be in the absence of globally co-ordinated action.

It is difficult to know whether to be more amazed at the insignificance of the warming that theoretically would be averted by a 5% cut in Australian emissions, or that politicians think it sensible to strive to attain such cuts, or that so many media reporters and commentators have failed to inform the Australian public of the relevant figures.

In any event, a direct cost of $127 billion by 2020 alone — as estimated by the federal Treasury, and that without taking flow-on increases in energy or fuel costs into account — for the first steps towards a notional warming averted by 2050 of between 0.00007°C and 0.0007°C seems unlikely to be greeted as a good idea by your average Aussie battler.

Why is Labor so certain that its carbon dioxide tax can't be repealed?
Because of the many financial and political poison pills that it has planted to inhibit repeal.

Labor's confidence that the Coalition will be unable to remove the carbon dioxide tax is a reflection of the time, effort and public money that they have spent over the last five years, supported by a host of special interests, to embed deep in the Australian psyche the idea that there is a need for such a tax. In the process the government has fed off or enlisted support for their policies from major environmental groups, large companies, the financial markets and any and all trade and industry associations, cheered on awhile by the major media outlets.

At some point in this courtship process, the directors of companies deemed to be 'major polluters' have had to consider their options and cover their bets. Individual companies were offered compensation carrots of up to a quarter

of a billion dollars or more, accompanied by the stick of penalties for non-compliance. It was extremely difficult for company board members to refuse such payments, and to stand against the compensation that the Labor Government was offering. To do that, directors would have to

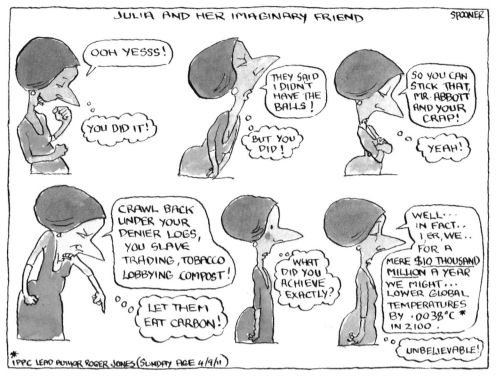

Prime minister Julia Gillard's ambiguous triumph in introducing
a carbon tax despite her promise not to. July 2012.

believe, and to be able to demonstrate to their shareholders, that their business
would be able to deliver financial outcomes in the future that outweigh the
bird-in-the-hand compensation on offer from the government.

At the moment, the carbon dioxide tax is, like a bureaucratic cancer, in the
process of metastasising throughout the Australian economy from foothold
centres of infection long ago established in every state (this is no surprise,
remembering that in 2007 Labor governments existed both nationally and
in every state).

Specifically, every state has conducted an audit of carbon dioxide emissions using one of the Big Four Accountants, followed by the implementation
of state-based carbon dioxide emissions systems. In NSW, trading certificates
were created and an Emissions Trading Certificate market was established. The
Independent Pricing and Regulatory Tribunal in NSW is currently charged
with undertaking a review of this market every year, together with summaris-
ing price movements and implementing penalties for defaulters — and this
system started as long ago as 2003.

A federal initiative by a new Coalition government to repeal the tax is therefore bound to create significant tensions with nearly all the states and territories.

Meanwhile, the Productivity Commission has gone quiet on the issue of pricing carbon dioxide. Undoubtedly, the Commission will benefit from lots of inquiries on this and related topics over the coming decade irrespective of who forms Australia's next government, and perhaps that is why it appears to have ducked for cover.

These are the reasons why Labor and the Greens are exuding such confidence that their penal carbon dioxide policies will survive into the future, even through a change of government. Talk about a climate change cuckoo in the government policy nest.

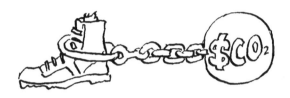

So can the Coalition keep its promise to unwind the tax?
Repealing the tax will be difficult, and expensive, but retaining it will cost 100's of billions of dollars

Since July 1, 2012, Australia has been committed to the taxation of carbon dioxide, and this demonisation of the gas will continue unless the Coalition wins government at the next election. The tax has become an all or nothing political commitment by the Labor-Green government.

However, the carbon dioxide tax is a complete reversal of the bipartisan industry policy that has been, mostly quietly, implemented across Australia over the last forty years. The direct cost of leaving the tax unchanged is estimated by Treasury to be $127 billion by 2020 alone, and the indirect costs and continuance beyond 2020 will multiply that figure several times over.

The Coalition has suggested that when and if in government it is unlikely to finance the present government's financial handouts, nor continue the programmes of social and economic change regarding global warming and climate change, all of which are the express antithesis of the Coalition's espoused bid for elevation to government. The next election is therefore foreshadowed to be about the core issue of whether the Australian people want to maintain a price penalty on carbon dioxide emissions, and should the Coalition win it will have committed irrevocably to unwinding the tax.

That the tax will be difficult to unwind has little to do with theory or philosophy, but is because the Labor-Green government has created an army of climate change organisations that are staffed largely by True Believers in positions of significant power and influence.

Therefore, multifarious procedures and processes will need to be cancelled or amended by a new Coalition government. Such changes will have to proceed in the face of endless self-interested arguments against repeal and policy reversion that will be advanced by state governments, by the major carbon dioxide emitters and by the large number of entrenched lobby organisations and industry associations that have led the carbon dioxide crusade in Australia. These networks, organisations and persons will be lined up in direct opposition to the position of a new, legitimately elected government which has said clearly and repetitively that it will repeal the carbon dioxide tax.

A new repealing government will also have to write off tens of billions of dollars in departmental studies, and make payments to the large number of companies and institutions that have come to believe, either willingly or under duress, that a carbon dioxide tax and its successor, an emissions trading scheme, were politically non-negotiable.

Removing this big, bad tax is therefore going to be far from easy. But countering the inefficient state schemes, and stopping the handouts to the advocates and rentseekers of the climate change industry, are all essential if the Australian economy is to prosper into the future. It is perhaps worth remembering that it has taken all the time since Federation to (just about) remove customs duties, and that sales tax was in place for nearly seventy years before it too was removed. Though the majority of the populace clearly approved of these actions, very few price reductions occurred as a result of the removal of these taxes.

So should you think it unreasonable to meet the high costs associated with removing the penalties on carbon dioxide, simply reflect that they are almost small change compared with the cost (and economic and social damage) of leaving the present and planned carbon dioxide tax-ETS system in place.

But aren't we doing all this for our children and grandchildren?
If we are, then we must be silly.

Those who lobby for environmental causes, and especially those who demonise carbon dioxide, often resort to emotional arguments. This is a reflection of the lack of substantive scientific evidence that carbon dioxide emissions are harmful; indeed, the evidence indicates that they are beneficial.

Prime amongst the emotional arguments is the one that surely all of us want to leave the environment in good shape for our descendants, who are going to inherit the planet. Of course, it is true that most citizens do indeed espouse this obviously good cause — hence the power of the argument.

But that the premise that we wish to nurture the environment is true does not justify the implied conclusion that the way to achieve the premise is to tax carbon dioxide!

History demonstrates that environmental preservation is delivered not by noble ambition but by economic wealth. For it is the wealthy western and OECD countries that spend by far the largest amounts of money on environmental matters, simply because they have the discretionary income to spend in the first place.

One of the keys to wealth generation, particularly in impoverished nations, is the provision of cheap energy, the availability of which acts both to improve the comfort of living and to encourage development and economic growth.

Exercising penal taxation against human-related carbon dioxide emissions, including the demonisation of cheap coal-fired power stations, therefore acts in precisely the wrong direction. By denying or increasing charges for power in under-developed nations, carbon dioxide taxes and trading schemes ensure that environmental and health damage continues unchecked, for example, by forcing people to use wood, charcoal or dried dung as a primary energy source. The taxes and trading schemes also inhibit much-needed development and wealth generation, and are thereby a direct source of unnecessary deaths.

In short, and as is indeed going to be only too apparent to your grandchildren, carbon dioxide taxes, trading schemes and similar policy options are not just economically damaging but immoral to boot.

XI

ALTERNATIVES TO ALTERNATIVE ENERGY

Electrons are clean: what on earth is all this about dirty energy?
Clean and dirty energy are figments of the emotional imagination.

Electricity, consisting of a stream of electrons,[47] is the cleanest and most versatile form of energy available to modern society. Electricity provides heat and light, and powers home appliances and industrial machines alike. Electric power is essential for computers, television, telephones and the internet, methods of entertainment and communication without which modern society and its governance would not be possible.

There is no such thing as 'clean' or 'dirty' energy. Instead, these terms really apply to the mechanisms by which industrial energy is generated. They are not scientific terms, but emotive words that are almost invariably used with a political intent.

[47.] An electron is a stable subatomic particle that is found in all atoms, and carries a negative charge. Flowing electrons act as the primary carrier of electricity.

All forms of electricity generation have an effect on the environment, and all societies try to minimise such impacts in a cost-effective way. For a coal-fired power station, the biggest impacts are ash and particulate emissions, which are filtered out at all modern generating stations. Wind farms are visually obtrusive, despoil the landscape, kill endangered eagles and other birds and bats, and cause serious medical problems because of low frequency noise. Solar farms require large areas of land to generate significant amounts of power, which again has a large visual and social impact. Nuclear power has no significant impact beyond that of that of manufacture, construction and decommissioning at the site of the power plant. Although widespread public fear exists about the disposal of radioactive waste, the technical problems of safe disposal are not great and recent research shows that low levels of radiation are not as dangerous as has been assumed. Lastly, hydropower creates significant landscape change (drowning of former river valleys, and the formation of lakes) which some people see as deleterious, although an alternative view is that as well as energy hydro-lakes provide both useful resources (stored water) and recreational value.

No objective ranking can be made of these various power sources from 'dirty' to 'clean', nor from environmentally 'friendly' to 'non-friendly', for such rankings differ with the location and are anyway subjective; judgement always lies very much in the eye of the beholder.

Contrary to the dominant view of environmentalists, the evidence shows with clarity that nuclear generation is the safest and most environmentally friendly form of industrial-scale power (below: Why isn't nuclear energy part of Australian and NZ planning?). For example, no one has died — or even become seriously ill — from radiation caused by the recent Fukushima accident in Japan; furthermore, because the radiation levels were low despite the very large scale of the disaster, it is unlikely that anyone will die in the future either.

The modern world depends entirely upon the availability of a reliable supply of electricity at a reasonable cost. Without electricity, modern society would collapse in a week, and within a few months millions of people would die. The differing environmental impacts of various potential energy sources therefore need to be assessed for social, environmental and cost-effectiveness on a case by case and location by location basis. Different implementation choices will be made by different societies and in different places, and one-size solutions most definitely do not fit all.

How much use is windpower?

Very useful in an isolated location: much too expensive to compete with conventional power generation to feed a national electricity grid.

Hundreds of years ago, animals, wind and water were the main suppliers of motive power. Watermills and windmills supplied a small amount of power from very large and expensive pieces of machinery — but only when sufficient water was available or when the wind blew. When the steam engine came along, these sources were abandoned despite the enormous expense of motive power from early steam engines.

Using a modern windmill to supply an isolated outback homestead with power often makes sense, but using a windfarm to supply a national energy grid does not. This is because windfarms produce an unpredictable and rapidly fluctuating output of electricity, require the expensive construction of new power transmission grids, cannot easily store energy generated for future or peak demand and require back up by a conventional coal or gas-fired power station during calm periods.

Without subsidies, wind farms cannot compete with conventional large scale power generation. In the global market, including USA and Europe, about 70% of the income of a windfarm comes from government-provided tax breaks and subsidies from consumers. In addition, the consumer also pays further hidden subsidies for new transmission lines and for the backup power generation that is needed when the wind stops blowing. Without subsidy, the real cost of windpower is about three times the cost of the same energy provided from a nuclear station and four times that provided from a modern coal-fired power station (compare Table 5).

In effect, wind power today suffers from the same problems as it did hundreds of years ago. Expensive machinery, a fluctuating supply driven by the vagaries of the wind, and a low average output of power.

Why aren't windfarms a cost-effective source of base-load electricity?

Because they provide only intermittent and expensive power

Wind farms cannot provide a steady supply of power because the wind does not blow all the time, and they are therefore cost-ineffective.

Windmills have proved to be not only publicly unpopular
but also environmentally dubious. October 2009.

The power from a wind turbine increases by a factor of eight if the wind speed doubles, and at wind speeds of more than about 90 km/hr turbines must be shut down to avoid damage. So windmills don't generate much power in light winds and they don't generate anything in strong winds. Typically, a wind turbine generates less than 10% of its rated power for 30% of the time, and more than 80% of its capacity for only about 5% of the time. The upshot of this is that a wind turbine rated at 1 MW will actually have an average output of about 0.3 MW, i.e. about one-third of its rated capacity.

The established capacity factor[48] of wind farms is about 35% in Australia and 37% in New Zealand. Worldwide, the average figure is lower, at around 25%. In February 2012, a month when southeast Australian demand for power was at its highest, the average wind output was in the region of 25-30%, exceeded 65% only once and was below 10% 12 times. At the time of the highest demand the output was 18%.

[48] The capacity factor of a wind farm is the ratio between the actual output during a year and the theoretical output that would have been generated if the wind farm had operated at full power for the whole year.

According to Meridian Energy, the 420 MW MacArthur wind farm in southwestern Victoria will cost a total of A$1 billion, which implies a rate of about A$2000/kW, compared to about $1500/kW for power from a modern coal-fired station. Based on a 20 year life, the overall cost of MacArthur generation then calculates to be about 12 cents/kWh.

But, remembering that the capacity of an Australian wind farm is about 35%, it typically takes 2500 MW of wind farm to generate the same amount of electricity as a 1000 MW coal-fired power station. This means that the capital cost of wind on an equal energy basis is actually $6600/kW, which is not cheap. The need for 2500 MW of wind power instead of 1000 MW of conventional generation also means that 2500 MW of transmission lines, transformers and the like are needed to transmit the power. All of this adds materially to the cost of electricity generated by wind farms, and is paid for by the consumer and not by the wind farm owner.

Australian power generation costs

Technology	Cost: c/kWh
Conventional black coal	7.5
Conventional brown coal	8.5
Combined cycle gas turbines	8.5
Nuclear	~12
Wind farm	18
Open cycle gas turbines	22
Photo voltaic solar	43

Table 5. Relative costs of power in Australia for different generation sources (after EPRI report to the Australian government, February, 2010.)

Wind power is expensive for other reasons as well. Wind farms cannot be guaranteed to produce a significant amount of power during peak demand periods, and in many cases wind power output during critical peak demands is in the region of 2-10% of maximum output. Therefore backup power stations must be built that are able to respond to the very rapid swings in output that occur from wind farms. The only feasible option is to use open cycle gas-fired turbine plants as backup, because only they have the capacity for fast response firing-up and shutting-down that is required. These power plants cost about $1000/kW to build, but are much less efficient than conventional coal-fired plant or combined cycle gas turbines, and have an all-in generating cost of about 20 cents/kWh. And once again, the consumer rather than the wind farm developer carries the cost.

Another problem with wind power is that, being produced at nature's whim, it cannot necessarily be used at the time of availability in the location where it has been generated. For example, Denmark has invested greatly in building wind farms, but about 60% of Danish wind generated power has to be exported to Scandinavia and Germany, because it cannot be used locally at times of low demand; however, when demand is high and the wind is not blowing in Denmark, then power has to be bought back from other countries on the European grid at a very much higher price.

Similarly, in New Zealand, the wind blows hardest in the springtime when the snow is melting, it is often raining and hydro dams are therefore full or filling. At such times an excess of wind power can occur, with the result that the power price collapses and hydropower stations are forced to spill energy. In Australia, the same thing often happens when demand is low in the early hours of the morning, thus forcing coal-fired stations to operate at reduced output (though with little reduction in fuel consumption or emissions). Because they often generate a lot of power when it is not needed, and little power when it is needed, the income for wind farms is always significantly lower than it is for power stations that are able to operate at full output during peak demand periods.

Lastly comes the issue of the service life of wind turbines. This is not well known from direct experience because the large units now in operation have only been running for a few years. However, very few engineers believe that wind turbines will last longer than about 20 years in operation, and there is also evidence that as they get older their maintenance costs and downtime start to increase rapidly. A late 2012 study of 3000 wind turbines by Scottish economist Professor Gordon Hughes found that decreasing turbine reliability occurs for machines older than 12 years, which then become uneconomic. In contrast, the normal life of a coal or gas-fired station is 30 to 40 years, for nuclear it is 60 years and hydropower stations can last for hundreds of years. The capacity factor for UK windfarms therefore drops from 25% to 15% in five years, and the decline is even more rapid for offshore wind farms in Denmark. The declining capacity factor with age makes wind farms even less economic than is usually assumed, because costing calculations usually assume that the capacity factor will not change over a 20 year lifespan.

Summing up, the reality is that windfarms are cost-ineffective; they only provide expensive and intermittent electricity at unpredictable times, and those times often do not coincide with the peak demand periods when reliable supply is vital.

But surely windfarms are environmentally beneficial?
It is a myth that wind power is environmentally beneficial.

For the last 20 years environmental activists, wind farm developers and other special interest lobby groups have contended remorselessly that the extra expense of wind farms is justified because they are more environmentally friendly than coal-fired power stations.

This assertion rests on two myths: the first is that coal-fired power stations pollute the atmosphere, which, with modern scrubbers fitted, they do not. The second myth is that wind farms are environmentally benign, which they most decidedly are not; the problems include noise pollution, social division, landscape desecration, killing bats and birds and, irony of ironies, a failure to substantially reduce net carbon dioxide emissions.

Wind farms occupy large areas of land, and because they are normally built on hills they are visually obtrusive. Major new roads are required to carry the large pieces of machinery to their final location, voluminous concrete foundations are needed, and extra space is needed for the massive cranes that are used during erection and maintenance.

For example, a standard 1000MW coal-fired power plant has a typical footprint less than one square kilometre in area. In contrast, a wind farm designed to generate the same amount of power would comprise 1700 x 1.5 MW windmills. As the recommended wind turbine separation is about 450 metres, siting the turbines in 3-wide rows will result in a wind farm more than 250 kilometres long and 100 metres wide (250 sq km). With the exclusion zone for housing being larger still, the final footprint of such a wind farm can occupy as much as 1,000 km².

The direct environmental footprint includes also the roading and site preparation works; the manufacture and transport of the concrete, plastic and steel involved in constructing each turbine; the construction of a back-up gas-fired power station; and finally the construction of the major transmission grid needed to link individual turbines to each other and then cross-country to the existing energy grid.

However, the damage does not stop there, because it is well established in many countries that wind turbine blades kill many birds and bats. The American Bird Conservancy estimates that by 2030 well over 1 million birds will be killed in the United States each year by wind turbines[49]; some of these are protected species, including eagles and hawks. Spain's Ornithological Society has estimated that the 18 thousand windmills in that country could be killing 6 million or more birds and bats every year.[50]

In summary, to replace a 1000 MW coal-fired power station requires a wind farm of 1700 large windmills that would effectively rule out any housing over an area of 1,000 km^2, plus a gas turbine power station of at least 500 MW nearby, all connected by new 400 kilovolt transmission lines and with attendant wildlife and human impacts.

The idea that all of this is somehow environmentally friendly is surely grotesque.

Well, at least windfarms save carbon dioxide emissions, don't they?

If at all, then not very much.

Atmospheric carbon dioxide, including that emitted as a result of human actions, is environmentally beneficial because it makes plants grow better. Nevertheless, governments have been persuaded by their IPCC-linked advisers that carbon dioxide is a pollutant and that building windfarms will significantly reduce industrial carbon dioxide emissions.

Many wind farms save some carbon dioxide compared with an equivalent coal-fired station, but not nearly as much as people have been encouraged to believe. The main reason is that the variable output of wind farms forces existing thermal power stations, which were designed for maximum efficiency whilst operating steadily at high outputs, to run over a wide and rapidly fluctuating range of conditions. As a result, these stations emit more ash, carbon dioxide, sulphur dioxide and nitrous oxide per unit of electrical output than they would if they were running under the conditions they were built for. Per

[49]. American Bird Conservancy's Policy Statement on Wind Energy and Bird-Smart Wind Guidelines. *http://www.abcbirds.org/abcprograms/policy/collisions/wind_policy.html.*

[50]. SEO/BirdLife presenta una nueva guía para la evaluación del impacto de parques eólicos en aves y murciélagos. *http://www.seo.org/2012/01/12/seobirdlife-presenta-una-nueva-guia-para-la-evaluacion-del-impacto-de-parques-eolicos-en-aves-y-murcielagos/*

unit of power output, the volume of both real pollutants and carbon dioxide emitted are increased. In addition, new open cycle gas turbine power stations needed to provide backup also add to the amount of extra carbon dioxide that is generated in association with the building of a wind farm.

Studies by Bryce Bentek for Texas and by Joseph Wheatley for Ireland show that the reduction in carbon dioxide caused by the introduction of a wind farm varies from zero to about two-thirds of the carbon dioxide emitted by alternative fossil-fuelled stations.

The same appears to be true in Australia, where a recent analysis by engineer Hamish Cumming, based on information from Australian electricity retailers, showed that the major coal-fired stations in Victoria reduce output but not coal consumption during times when wind power becomes available. So, although the availability of wind power does allow the stations to reduce their output, there is no corresponding reduction in carbon dioxide emissions — this is surely quixotic environmental policy.

What about solar power, then?
Solar power is expensive, and cannot run a national electricity grid.

Solar power is about three times as expensive again as wind power, and the capacity factor varies from 9% for well-sited installations in Germany to about 22% in a desert. Typical capacity factors in a sunny country are about 18-20%.

Solar power disappears every night when the Sun goes down. It also drops by up to 60% if a cloud obscures the Sun. As a result, standby power plant is always needed for solar installations, and, as for wind farms, the most feasible backup is the construction of an open cycle gas turbine power plant. In most countries peak electricity demand occurs either on winter evenings (temperate latitudes) or in the afternoon of hot days (tropical latitudes). On winter evenings solar output is of course virtually zero, and from about 2.00 pm onwards solar output drops rapidly even in the tropics. By 5.00 pm, which is a typical time for a tropical peak demand to occur, solar output will have dropped to about 50% of its maximum.

As with wind power, solar power seldom produces its rated output. This can be because of dust on the solar panels, because the panels are not properly aligned towards the Sun or because the panels have aged. Because of their lower capacity factor than wind turbines, 4000MW or 5000 MW of solar installation is needed to produce the same energy as a 1000 MW conventional station.

Solar power has all the same problems of transmission, backup and additional emissions from backup conventional power stations that we have

already discussed for wind power (above: Why aren't windfarms a cost-effective source of base-load electricity?).

Large scale solar power generation exists only because, pressured by green lobby groups and renewable energy developers, governments have decided to bestow large taxpayer and consumer subsidies upon it. Without subsidies, solar power would be used only where electricity is needed in a location too remote to be supplied by the main grid. Small solar panels are indeed a convenient and cost-effective way to supply small electricity demands in remote places. But until the price drops, and a way is devised to store electrical energy cheaply for periods of weeks and months, solar will remain a long way away from being able to produce cost-effective grid electricity.

Perhaps tidal power is the solution?
In principle, a large source of power: in practice, cost-ineffective and unreliable

Another apparently attractive source of power are the tides. Huge quantities of water move in a regular and predictable way during the daily tidal cycle, and particular geographies can concentrate the flows into regions of high power potential. Tidal power comes in two forms: schemes that use a barrage across a river or estuary, and those that rely on tidal streams passing through the marine equivalent of a seabed windmill.

Barrage generators
A tidal barrage takes the form of a dam built across an inlet where the tidal range is very high — usually more than eight metres. Tidal barrage schemes were first developed more than 1000 years ago, and about 800 years ago there were 76 tidal mills in London. These mills were superseded by steam engines.

Modern barrages containing water turbines were first constructed in the 1960s, an early example being the 1.7 MW Kislaya Guba Station in Russia. In 1966, Electricité de France (EDF) developed the larger 240 MW La Rance scheme in Brittany. No costings have been made available for these stations, so it can be assumed that the price was embarrassingly high. No other tidal power schemes have been developed since in Europe, although many British governments over the last 70 years have toyed with the idea of building a barrage over the River Severn estuary; so far the high costs, which are well above those for a nuclear station, have defeated the scheme.

The only recent tidal power project is at Sihwa in Korea. At 250MW, the scheme is slightly larger than La Rance, and has been based on an existing barrage that was built to form a freshwater lake. When the lake became polluted, the barrage was breached to flush the lake with seawater. It was then decided to build a tidal power scheme. Because it did not include the cost of the barrage and other works, the apparent cost of 'only' $300 million ($1,200/kW) appears to be reasonable. However, given the capacity factor of about 25%, the like-for-like comparison with a nuclear power station operating at 90% capacity factor gives an equivalent cost of $4,300/kW, on top of which backup plant is also needed. The scheme would have clearly been uneconomic if its financing had had to carry the construction of the barrage.

In 1998, co-author Bryan Leyland worked on a tidal power scheme for Derby in the Northern Territory of Australia. Although the scheme had many advantages, such as two adjacent creeks to form upper and lower basins, an 8.5 metre tidal range and subsidies from the government to cover the very high cost of diesel generated power in Derby, the operation was still too expensive to be profitable and has never been constructed.

Barrage power schemes suffer from the high costs of low-head generating plant and high civil engineering costs for the barrage, the powerhouse and the various control gates, combined with a fluctuating power output. Another major problem is that tidal stations generate power for only about eight hours out of every 24, most operating only when the tide is ebbing. This means that equivalent conventional generating capacity has to be held in reserve on all occasions, and used to replace the tidal station output at times when the tidal output is zero yet demand may be high. Like windfarms, because they require such expensive backup generation tidal power stations do not add any firm capacity to a national grid system. There are relatively few suitable sites, and that each project is inevitably unique adds to the design costs.

The very high initial capital costs of constructing a barrage, and long delay before financial returns are generated, makes it difficult to attract private investment into them. It is not easy to see how these problems can be overcome to the extent that tidal power could compete with conventional generation sources.

Tidal stream generators

Tidal stream generation uses shallow water tidal currents to generate electricity in the same way as a wind turbine does. They generate power intermittently and they must be engineered so as to survive in an aggressive environment with strong currents in two directions. As a result, tidal stream generators tend to be heavy, and this means that they are inherently expensive.

The ready availability of subsidies and grants in Europe and the USA has spawned many interesting concepts for tidal stream generation. Some prototypes have been tested and have a capacity factor of around 25%. But as for conventional tidal barrage power generation, they operate for only a few hours each day. They also generate much less during neap[51] tides, because like windmills the turbine power follows a cube law.

In conclusion, tidal power schemes based on barrages use a well-developed technology, but that technology is not cost-competitive with conventional hydrocarbon-based power generation. Tidal current turbines are a developing technology with particular potential for island communities, but because of their weight and intermittent operation they too are currently uneconomic and uncompetitive.

How do biofuels benefit the environment?
They don't; and nor does using them as an energy source help us to meet the world's food supply

The derivation of alcohol (bioethanol) or hydrocarbon-based (biodiesel) fuels from plant material has a long history that predates the global warming alarm of the late 20th century. Material that is used ranges from forestry waste to industrially planted sugarcane (bagasse), various native grasses, hemp, corn, sorghum and a variety of tree species including eucalypts and oil palms.

Bioethanol can be used as a replacement for, or blended into, petrol; and biodiesel is a replacement for conventional diesel.

Arguments in favour of using biofuels include the assertion that we need to move now towards renewable fuels because fossil fuels will soon run out. The assumption is that we may have already moved past 'peak oil' — the point at which we are using oil at a greater rate than we are discovering new supplies. It is also asserted that replacing fossil fuel with biofuel will reduce

---

[51.] The tidal cycle of 28 days exhibits two periods each of larger (spring) and smaller (neap) tides, with intermediate height tides occurring in between. Spring tides occur when the moon and Sun are in line, which causes their gravitational pulls to add to one another and exert maximum force upon the Earth. Neap tides occur when the Sun and moon are at right angles to one another; their gravitational pulls then partially cancel each other out resulting in a lesser net tidal force being applied to the Earth.

emissions of carbon dioxide. And a third argument is that home-grown biofuel can generate employment and wealth, provide energy diversity and security and act as a cushion against arbitrary price rises for imported fuel.

The first two of these arguments are almost completely fallacious. Regarding peak oil, the idea that we have already reached, or are poised to attain, maximum use of hydrocarbon resources has been comprehensively discredited. New technologies (including horizontal drilling and reservoir fracking) now allow dispersed oil and gas to be recovered at relatively low cost from the very large reserves that occur in fine-grained and often 'tight' sedimentary rocks (shale, coal). Meanwhile, huge, yet-to-be-tapped energy resources occur as methane gas-hydrate deposits beneath the sea-bed along continental margins into which preliminary exploration testing wells have already been drilled. Based on these resources, conventional hydrocarbon energy supplies are likely to remain available for hundreds of years into the future.

Second, the argument that growing biofuel will cause a decrease in greenhouse emissions is seldom true. Independent studies have shown that once account is taken of the emissions from the forests cleared and burned to make way for monoculture plantations (with attendant loss of biodiversity), and the fuel burned and nitrous oxides released in planting, tending, harvesting, transporting and processing the crop, the actual result is often an overall increase in greenhouse emissions rather than the intended decrease. It is also reported that, as a result of blending biofuel into petrol, fuel consumption increases.

The third, and perhaps greatest, disadvantage of turning productive arable land over to biofuel production is that it reduces world food supply. With world population growing towards an estimated total of around 10 billion, and given that most of the great grain-growing areas are located in cold-temperate latitudes and therefore vulnerable to even a minor decline in temperature, using agricultural land or cutting down forests to grow fuel rather than food crops is clearly an unwise policy option.

In addition to this, the aggressive planting of cereal crops for biofuel can cause large increases in the price of corn and other grains. This occurred in 2007–2008 when widespread droughts coincided with an increase in government subsidies for turning food crops into biofuel caused large and widespread increases in food prices. In December 2007, the UN Food and Agriculture Organisation estimated that world prices of sugarcane, corn, rapeseed oil, palm oil, and soybeans had risen 40% in the preceding 12 months. The resulting soaring cost of food led to riots and unrest in the parts of Africa, the Middle East and Latin America that rely upon imported food. Meanwhile, in Malaysia and Indonesia, where large palm oil plantations had been established after rainforest clearances, biodiesel refining created a local palm oil shortage for cooking, with price increases up to 70%.

Ignoring both science and economics, biofuel production throughout the world has been stimulated this century by strong government tax incentives and subsidies. In response, ethanol production (mainly from USA and Brazil) tripled from 4.9 to almost 15.9 billion gallons between 2001 and 2007, and over the same period biodiesel production (a favourite in the European Union) rose almost tenfold, to about 2.4 billion gallons.

Noting these events, the Head of the Earth Policy Institute, Lester Brown, commented:

> We are witnessing the beginning of one of the great tragedies of history. The United States, in a misguided effort to reduce its oil insecurity by converting grain into fuel for cars, is generating global food insecurity on a scale never seen before.

The very high environmental and economic costs of biofuels means that even the argument that a diversity of suppliers or sources is a good thing for any commodity of national importance (which liquid fuels represent) lacks merit. Consequently, the environmental and social damage caused by growing biofuels needs to be stopped as soon as possible. In which regard it is difficult to disagree with the opinion of Peter Brabeck-Letmathe, chief executive of Nestlé, the world's largest food and beverage company, who in March 2008 said:

> If, as predicted, we look to use biofuels to satisfy 20% of the growing demand for oil products, there will be nothing left to eat. To grant enormous subsidies for biofuel production is morally unacceptable and irresponsible.

Why isn't nuclear energy part of Australian and New Zealand planning?
Because of public fear accompanied by a lack of urgent need.

Contrary to public perception, neither Australian law nor New Zealand's anti-nuclear legislation prohibit the construction and operation of nuclear power stations. Nonetheless, a majority of people in both countries have been led to believe that nuclear power is horribly unsafe, and many are now strongly opposed to its development. The reality is, however, that both countries are so rich in other low-cost energy resources that no urgent necessity exists to develop nuclear stations for at least several decades.

Australia has very large amounts of coal (which generates about 80% of the current energy supply), bountiful conventional and unconventional gas, and not insignificant reserves oil; and New Zealand has large hydropower resources (about 70% of current supply), large amounts of coal and strong potential for large new finds of onshore and offshore gas. In addition, recent energy demand has been blunted in both countries; in Australia by a slackening of growth and a pause in the mining boom; and in New Zealand because of the Christchurch earthquake and a slackening economy which has included the the closure of timber plants because of high electricity prices and emissions trading costs.

Looking to the long-term future, little doubt exists that nuclear power will play an increasing part in energy production worldwide, because it is by a large margin the most environmentally friendly and safest of all forms of power generation. Despite the widespread belief that nuclear power is dangerous, the death rate from nuclear accidents has, in fact, been extremely low. The United Nations has estimated that fewer than 50 persons died during the Chernobyl disaster, despite the reactor there being of an obsolete design and failing to have proper shielding. Including Chernobyl, only 48 deaths have occurred in association with nuclear power plant operations since 1969 (Table 6). In contrast, coal-fired power stations are responsible for the deaths of thousands of miners worldwide each year, and hydropower stations also have the potential to be extremely dangerous; for example, the 1975 failure of the Banqiao Reservoir Dam in Henan Province, China, killed an estimated 171,000 people and 11 million people lost their homes.

Recent research by the United Nations Scientific Committee on the Effects of Atomic Radiation shows that nuclear radiation is less dangerous than current regulations assume. Were the regulations to be adjusted to match reality, both the fear of nuclear power and its cost would diminish. Meanwhile, neither Australia nor New Zealand have any urgent need to develop nuclear power stations until the new generation of modern stations being developed overseas are in commercial operation. After settling into serial production, and becoming available at a competitive cost, these modern nuclear power stations are certain to become a feasible, clean and cost-effective power source with sufficient fuel for at least many hundreds of years.

Beyond that again, thorium reactors and nuclear fusion continue to beckon as power sources of the future.

Energy source	Number of accidents	Total fatalities
Coal	1221	25,107
Oil	397	20,218
Natural gas	135	2,043
LPG	105	3,921
Hydropower	11	29,938
Nuclear	1	31
Total	1,870	81,258

Table 6. Relative hazard of different types of electricity generation. (from *http://gabe.web.psi.ch/pdfs/_2012_LEA_Audit/TA01.pdf*)

What's wrong with coal-fired power stations, anyway?
Nothing; in reality they are environmentally beneficial

There is nothing wrong with coal-fired power stations. Modern stations have minimal emissions of dust, soot and harmful gases such as sulphur dioxide or nitrous oxide because these genuine pollutants are removed by scrubbers before they get to the smokestack. The main environmental impacts are the need for cooling water and the need for large areas to store the ash that results from burning the coal.

Furthermore, coal is a widespread commodity and cheap, so coal-fired power generation remains one of the least costly ways to generate large amounts of power. Coal is therefore a godsend for developing nations, which tend to be both energy and money poor. To deny poor people the right

to utilise coal-fired energy to develop their economies, as many environmental lobbyists want to do, is unacceptable. Indeed, one American writer has gone so far as to label such actions technological genocide — drawing parallels with the similarly misguided ban on the use of DDT that was only recently lifted by the United Nations. For, beyond a shadow of doubt, both these policies result in increases poverty and mortality in third world nations.

As explained in IV: Is atmospheric carbon dioxide a pollutant?, man-made carbon dioxide emissions do not cause dangerous global warming, but do enhance plant growth, including stimulating efficient water use, for double benefit. Emissions therefore help to green the planet and feed the world. For instance, recent studies estimate that between 1989 and 2009 about 300,000 km^2 of new vegetation became established across the African Sahel region, in parallel with the increasing levels of carbon dioxide in the atmosphere. Far from being a problem, then, carbon dioxide emissions are environmentally beneficial. And increasing emissions by burning coal yields the extra benefit of the provision of cheap power for all nations.

The good news in the most recent BP Statistical Review is that proven world reserves of coal are sufficient to meet 112 years of production at current levels, and that the consumption of coal increased by 5.4% in 2011. Against this background, the World Resources Institute reported recently that 1,231 new coal plants with a total capacity of 1,400 GW are scheduled to be built worldwide, including 79 in the USA; for comparison, Australia currently generates 35 GW of coal-fired power.

New coal plants are not only being constructed in developing nations. In Germany, a country with extremely influential Green politicians, power utility companies have recently announced plans for the construction of 25 new coal-fired plants. These stations are needed to avoid a looming power shortage engendered by the fatal combination of favouring wind turbine and solar cell construction and decommissioning nuclear power plants that have operated safely for many years. Many German coal-fired plants burn brown coal (lignite) which produces the highest emission levels of any hydrocarbon-based fuel. That more than 25% of Germany's energy, and growing, is now produced from plants of this type is surely going to be much appreciated by German farmers and market gardeners. For comparison, again, in Australia the brown coal power plants of the La Trobe Valley currently provide 22% of the nation's power.

Doubtless the new German policy of coal power plant construction has been informed by the devastating realisation that the country's 30-year long love affair with alternative energy sources has, at very great cost, weakened the reliability of the energy grid and ended up damaging rather than improving the environment. For, as recently summed up by the archetypal

---

[52.] Germany's 'Energiewende' – the story so far. *http://www.marklynas.org/2013/01/germanys-energiewende-the-story-so-far/*

environmental writer Mark Lynas[52]:

> *Unfortunately, Germany's 'renewables revolution' is at best making no difference to the country's carbon (sic) emissions, and at worst pushing them marginally upwards. Thus, tens (or even hundreds, depending on who you believe) of billions of euros are being spent on expensive solar PV and wind installations for no climatic benefit whatsoever.*

Politicians such as Green's leader Bob Brown persist in referring to the physical absurdity of 'decarbonisation'. May 2011.

XII

THE WAY FORWARD: PRECAUTION AGAINST NATURAL HAZARD

Surely we should give Earth 'the benefit of the doubt' about global warming?

Feel-good precaution won't protect us from the ravages of climatic events and change; hard-nosed and effective prudence will.

The precautionary principle was introduced to assist governments and peoples with the risk analysis of environmental issues. First formulated at a United Nations environment conference at Rio de Janiero in 1992, it states:

> *Where there are threats of serious or irreversible damage, lack of full scientific certainty shall not be used as a reason for postponing cost-effective measures to prevent environmental degradation.*

Faced as they are with a lack of compelling science on their side, many global warming activists invoke the precautionary principle as a means of forcing action against what they feel, but cannot show, is a strong risk of dangerous human-caused warming.

First, the very introduction of the precautionary principle into the argument in the first place is an acknowledgement that no compelling scientific evidence for alarm exists.

Second, the precautionary principle often represents a moral precept masquerading as a scientific one. This is a principle of the wrong type to be used for the formulation of effective public policy, which instead needs to be rooted in evidence-based science. Scientific principles acknowledge the supremacy of experiment and observation, and do not bow to untestable moral propositions or political fixes.

Third, if we wish to take precautions, we need to know what to take them against. Different computer models give different projections of future temperature, which means that no scientist can say with confidence whether the temperature in 2020, let alone 2100, will be warmer or cooler than today's. So do we take precautions against future warming or cooling; and which would be worse?

Fourth, those who propose the curtailment of human carbon dioxide emissions as a precaution against future warming invariably fail to address the critical cost/benefit issue — and this against the background that there is no credible published research that shows that the human costs and risks of a given amount of future global warming will exceed the costs and risks of an equivalent global cooling. For those asserting dangerous warming, the key cost/benefit question concerns what amount of 'warming prevented' will result from proposed emissions reductions schemes (X: How much warming will be averted by cutting Australian emissions?). The answer, for Australia, is that curtailing ALL industrial emissions would notionally prevent only a tiny 0.02°C of warming. This is a trivial amount, yet the costs of even attaining a small part of such warming averted (by, say, cutting emissions by 50%) will be in the multi-billion dollar range.

The slogan 'the benefit of the doubt' is deliberately emotional and bears all the hallmarks of having been produced by a green advertising agency. The catchy phrase reveals a profound misunderstanding of the real climatic risks faced by our societies, because it assumes that global warming is more dangerous, or more to be feared, than is global cooling. In reality, the converse is likely to be true.

What has climate change got to do with energy supply anyway?
Almost nothing.

It is a remarkable fact that virtually all governments now view climate change and energy supply as closely related policy issues. But hang on a moment: climate change issues are concerned with environmental hazards, whereas energy policy is concerned with supplying cheap, reliable and secure

German Chancellor Angela Merkel inspects a wind farm. All windmills must have a continuous coal or gas-fired power station back up for when the wind drops or gets too strong. November 2011.

electricity supplies to industry and the populace. Where is the relationship?

The answer is that until the 1980s there was no relationship, and that one is perceived now testifies only to the effectiveness of relentless lobbying by environmentalists, NGOs and commercial special interests towards the cause of connecting climate and energy policies. Truth, scientific balance and commonsense have been casualties along the way.

The conflation has been brought about by evangelising the view that carbon dioxide emissions from power generation using hydrocarbon-based fuels will cause dangerous global warming. That (false) view has become embedded in society to the point where now even prime ministers and presidents misuse 'carbon' as a shorthand for 'carbon dioxide', and then label it as a pollutant to boot (IV: Is atmospheric carbon dioxide a pollutant?).

As we have demonstrated earlier, carbon dioxide is environmentally beneficial; it is the elixir of life for most of our planetary ecosystems, and to badge it as a pollutant is therefore grotesque rather than just wrong. Second, the amount of carbon dioxide produced by human industrial processes is small compared with existing natural fluxes through the atmosphere and ocean (human emissions being less than 5% of natural emissions, see X: What percentage of carbon dioxide does Australian society generate?). Third, and most important of all, despite carbon dioxide being a greenhouse gas no evidence exists that the amount humans have added to the atmosphere is producing dangerous warming; or, indeed, any measurable warming at all.

Many negative consequences flow from conflating the energy and global warming issues, but foremost amongst them has been a lemming-like rush by governments to massively subsidise what are otherwise uneconomic sources of power — especially solar and wind power generation. These alternative sources are painted by lobby groups and governments alike as environmentally virtuous, because they are claimed to reduce carbon dioxide emissions as well as being both 'renewable' and 'clean' sources of energy.

Well, yes, wind and solar energy are indeed renewable when the wind blows and the Sun shines, but they are absent otherwise and tough luck if that is when you want to boil the kettle.

As we have seen (XI: Why aren't wind farms a cost-effective source of base-load electricity? and XI: Well, at least windfarms save carbon dioxide emissions, don't they?), both forms of generation are very expensive and their intermittency makes them unsuitable to be major contributors to a national energy grid. On top of that, the life of wind farm is about 15 years against 60 years from a nuclear station. In addition to their expense and impracticality, the claims as to the cleanliness and environmental friendliness of both solar and wind power generation are routinely overstated to the point of propagandisation.

Driven by environmentalists, and others who hope to make fat profits from carbon dioxide trading, a wilful increase in the cost and complexity of energy supply systems has occurred worldwide over the last two decades, and that for a negative environmental and social return. Amongst other symptoms, power prices have escalated sharply, and the situation has now become both economically and politically unsustainable. Nowhere has this happened to a greater degree than in Australia, which once had the cheapest electricity in the industrialised world; remember that?

As the political pressures build, so even the European Union is being forced to confront reality. For example, EU Energy Commissioner Gunther Oettinger recently stated in Berlin that European energy policy must change from being climate driven to being driven by the needs of industry.

What, then, needs to be done to improve the situation?

Individual nations must return to the formerly clear separation that they recognised between energy policy and climate policy, and analyse and plan for each with respect to their own separate requirements and resources. This means abandoning the woolly conflation of the two that has been so skilfully foisted on society by powerful vested interests over the last three decades. It also entails abandoning the monopoly IPCC advice about global warming and the use of fossil fuels, advice that engendered much of the confusion in the first place and continues to do so.

What are Australia's greatest natural hazards?
Nearly all are climate-related, including drought, bushfire, storms and floods.

Australia is one of the world's largest island continents. It is also one of the few large landmasses that lacks a geological plate boundary within its borders. Instead, emergent Australia lies within what is called the Indo-Australian plate of the global crustal jigsaw. To the north and east the boundaries of this plate lie along the mountainous terrains of the Himalaya and the Indonesia-Papua New Guinea and New Zealand volcanic arcs. To the south, the plate boundary corresponds with the submarine volcanic mid-ocean ridge that runs east-west through the Southern Ocean approximately midway between Australia and Antarctica.

In the absence of the destructive volcanic and large earthquake hazards that are associated with plate boundary tectonics[53], and aside from tsunami-risk, Australia's greatest environmental hazards are all climate-related. Droughts, floods, cyclones and large bushfires — Australia has them all in spades.

It must be recognised that the theoretical hazard of dangerous human-caused global warming is but one small part of a much wider climate hazard that scientists agree upon, which is the dangerous natural weather and climatic events that nature intermittently presents us with — and always will. It

---

[53.] Tectonics: the study of the deformation of Earth's crustal layers, and of the physical forces and geological processes that build continents, ocean basins and mountains, in the process bringing the deformation about.

is absolutely clear from, for example, the 2005 Hurricane Katrina disaster in the US, the 2007 floods in the United Kingdom and the tragic bushfires in Victoria, Australia in 2009, that the governments of even advanced, wealthy countries are often inadequately prepared for climate-related disasters of natural origin. We need to do better.

What does the climatic future really hold?

No one knows with certainty, but here are two alternatives to IPCC's dangerous AGW hypothesis.

It is now firmly established that any effect of human-related carbon dioxide emissions is, at most, very small, and perhaps even unmeasurably so. The best tool with which to assess speculative long-term climate change is therefore not the IPCC's unvalidated computer models, but rather an approach based on the projection of existing major natural trends and cyclicities. Let us look at two such examples.

One view of the future has been prepared by Japanese Professor Syun-Ichi

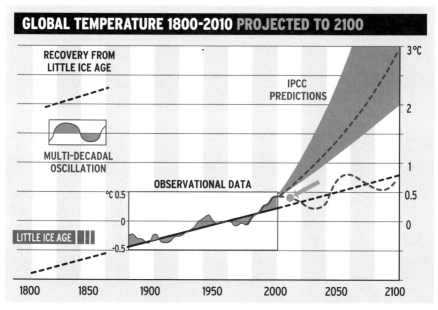

Fig. 42. Reconstructed global surface temperature, 1800-2100 (Akasofu, 2009). Observational data are represented within the central box. The modelled extension of global temperature beyond 2000 is based upon a continuing warming recovery from the Little Ice Age at a rate of 0.5° C/century, as modulated by a 60 year-long multi-decadal rhythm. IPCC computer projections are indicated by the pink field with central red dashed line. The arrowed green spot indicates the temperature in 2008, which falls on Akosofu's projected curve but beneath the field of IPCC predictions.

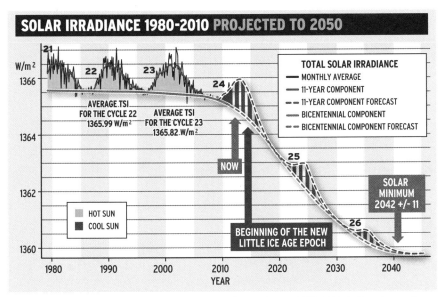

Fig. 43. Summary of observed variations in solar irradiance, 1979-2010 (yellow, hot sun), with projected future variations out to 2050 (dark red stripes, cool sun) (Abdussamatov, 2012). Solar cycles 21-26, numbered. The forward projection of historical trends and cycles indicates a near-future cooling of 6 W/m², culminating in a new Little Ice Age and solar minimum in 2042. This decrease in inbound solar energy, if it occurs, will be equivalent to a global cooling of about 1° C.

Akosofu (Fig. 42, p.232). Akosofu's projection is based upon the continuation of two natural trends: first the gentle linear warming of about 0.7°C/ century that has so far represented recovery from the Little Ice Age; and, second, modulating this trend, the well established multi-decadal cyclicity of about 0.5°C magnitude and 60–70 years duration that is represented by the PDO. This conservative analysis stands in stark contrast to the IPCC's alarmist computer projections (indicated in pink), but even the modest warming that Akosofu predicts may turn out to be an overestimate. First, because at some future stage the warming recovery from the Little Ice Age will terminate; and, second, because in due course, whether at the same time or otherwise, a long-term cooling decline will commence towards the next major glacial period.

A second view of the future, based upon the analysis of solar cycles, but allowing also for a human greenhouse effect, envisages near-term climatic stasis rather than a continuation of the post-Little Ice Age warming trend (Nicola Scafetta, Fig. 21, p.123). In a similar analysis but restricted to natural factors only, Professor Habibullo Abdussamatov from St Petersburg has analysed historical changes in solar irradiance and projected the pattern that they display out to 2050 (Fig. 43, p.233). The key prediction he makes, which has severe economic and social implications, is that we are about to enter into a new Little Ice Age epoch, with a Solar Minimum occurring near 2042 AD. Professors Scafetta and Abdussamatov are far from the only solar astrophysicists who worry about impending cooling and the disastrous world food shortages that could result.

Though these alternative futures are based upon relevant and good quality empirical data, no scientist knows if these, or any other empirical projection, will turn out to match future climate. Given the scientific uncertainty, however, it is premature to be continuing to implement anti-carbon dioxide measures based upon the unvalidated and inaccurate deterministic computer models of the IPCC, which address a hypothetical AGW that hasn't yet been measured let alone shown to be dangerous.

The only rational conclusion is that we need to be prepared to react to either warming or cooling over the next several decades, by developing science-based adaptive response strategies that can deal with whatever nature in the end serves up to us.

Do we really need a national climate policy, then?
You bet: for climate events and change are Australia's greatest natural hazards.

Given, then, that we cannot predict what future climate will be, do we still need a national climate policy?

Indeed we do, for a primary government duty of care is to protect the citizenry and the environment from the ravages of natural climate events. What is needed is not unnecessary and penal measures against carbon dioxide emissions, but instead a prudent and cost-effective policy of preparation for, and response to, all climatic events and hazards as and when they develop.

Climate hazard is both a geological and meteorological issue. Geological hazards are mostly dealt with by providing civil defence authorities and the public with accurate, evidence-based information regarding events such as earthquakes, volcanic eruptions, tsunamis, storms and floods (which represent climatic as well as weather events), and by mitigating and adapting to the effects when an event occurs. New Zealand's GeoNet[54] natural hazard network is a world-best-practice example of how to proceed. The additional risk of longer-term climate change, which GeoNet currently doesn't cover, differs from most other natural hazards only in that it occurs over periods of decades to hundreds or thousands of years. This difference is not one of kind, and neither should be our response planning.

The appropriate response to climate hazard, then, is national policies based on preparing for and adapting to all climate events as and when they happen, and irrespective of their presumed cause. Every country needs to develop its own understanding of, and plans to cope with, the unique combination of climate hazards that apply to it alone. The planned responses should be based upon adaptation, with mitigation where appropriate to cushion citizens who are affected in an undesirable or uncontrollable way.

The idea that there can be a one-size-fits-all global solution to deal with just one aspect of future (theoretical) climate hazard, as recommended by the IPCC, fails entirely to deal with the real climate and climate-related hazards to which we are all exposed. In their recent book, *Adaptive Governance and Climate Change*, Ronald Brunner and Amanda Lynch advance an argument supported by many scientists:

> *We need to use adaptive governance to produce response programs that cope with hazardous climate events as they happen, and that encourage diversity and innovation in the search for solutions. In such a fashion, the highly contentious 'global warming' problem can be recast into an issue in which every culture and community around the world has an inherent interest.*

What can I do to help achieve a sensible national climate policy?
Educate yourself in, and explain to others, the real facts of the matter.

Many people believe that the solution to reducing carbon dioxide emissions lies in making large changes in the behaviour of the average citizen. For example, Professor Kevin Anderson, who is Head of Climate & Energy Research, Tyndall Centre, Manchester University, recently wrote the

[54] GeoNet is New Zealand's national natural hazard monitoring agency. GeoNet operates networks of geophysical instruments to detect, analyse and respond to earthquakes, volcanic activity, landslides and tsunami.

following on the topic:

> *So what will I do differently? I haven't flown for almost eight years — and that will have to continue. I have halved the distance I drive each year and have significantly changed how I drive. I've done without a fridge for 12 years, but recently relented and joined the very small proportion of the world's population that has a fridge — this I may have to reverse! I've cut back on washing and showering — but only to levels that were the norm just a few years back. All this is a start but it is not enough. Certainly, if those of us working on climate change are a bellwether of society's response, the future looks bleak. Nevertheless until those intimately engaged in climate change, including the scientists, journalists, NGOs and ministers, put their own houses in order, I think it unlikely others will take our analysis seriously. As we pass the bus stop to jump in a taxi from the airport to another air-conditioned hotel room in Bali, Cancun or Rio — what message are we disseminating?*

Well, good luck with setting that brave example, Kevin. The reality is that there is no chance at all that such practices will be followed by the great majority of western citizens who currently enjoy the privileges that you are choosing to deny yourself. And that not least because it is becoming ever more widely understood amongst the populace that the underlying premise for Professor Anderson's actions — that carbon dioxide emissions are environmentally harmful — is false.

Two other things need to be said, however. The first is that setting national policies on issues such as climate hazard is the task of national governments, not the citizenry. And the second is the fact that man-made carbon dioxide emissions are beneficial rather than harmful does NOT mean that governments should ignore the genuine weather and climate hazards that exist naturally.

As discussed in response to the previous question (Do we really need a national climate policy?), the major climate-related hazards in Australia and New Zealand include cyclones, storms, floods, landslides, droughts and bushfires, as well as the longer term hazards of abrupt or extended warmings or coolings in temperature and their attendant consequences. We should deal with all of these matters by preparing for adverse climatic events in general, and by mitigating and adapting to the effects of individual events as and when they occur.

So, what can an individual citizen do towards achieving that end?

First, ensure that you are well briefed about the factual evidence that applies to the global warming and climate change issues. Simple facts are the best, such as that there has been no global warming since 1997 (16 years) despite an increase in atmospheric carbon dioxide of about 8% — so where is the crisis? Other similarly relevant facts can be garnered from throughout this book, or from elsewhere, always bearing in mind that projections from computer models do not constitute scientific evidence; instead, they represent more or less intelligent speculations about what might or might not eventuate in the distant future.

Second, and thus armed, work as hard as you can to share your understanding with your friends, colleagues and elected representatives. Along the way it will do no harm, either, to promulgate your evidence-based views on the issue through any media outlet that is open to publishing or broadcasting them.

Never forget that the idea that carbon dioxide emissions are causing dangerous global warming remains an unproven, and probably unlikely, scientific hypothesis. It must therefore continue to be tested against scientific facts.

In contrast, it is a fact rather than a hypothesis that natural weather and climatic events present deadly human and environmental risks. Pursuing expensive and futile schemes to combat the speculative, and quite possibly illusory, risks of human-induced global warming is both pointless and wealth-sapping. Instead, any sensible national climate policy must primarily address the well known risks of natural climate events and change.

Glossary

This brief glossary provides a summary explanation of (i) scientific measures as used throughout the text; and (ii) common acronyms and abbreviations. Other technical terms are defined or explained in the text or in footnotes at the point of their introduction into the text.

Scientific measures

Energy

kW, MW, GW   Domestic and industrial amounts of energy are usually expressed in terms of thousands (kW; for kilowatt = 10^3 watts), millions (MW; for megawatts = 10^6 watts), or billions (GW; for gigawatts = 10^9 watts) of watts.

W   The watt (W) is the basic unit used to measure electrical power, and is defined as a power rate of one joule per second. In turn, a joule (J) is a unit of energy, defined as the energy expended in moving 1 newton (N) a distance of 1 metre (m). Finally, a newton is the force required to accelerate a mass of one kilogram (kg) at a rate of one metre per second squared.

W/m^2 The amount of energy in watts received over an area of one square metre (watts/ square metre).

Length

μ   1 micron (μ) = one one-millionth of a metre m, 0.000001 m, 10^{-6} m (or 10^{-3} mm).

nm   1 nanometre (nm) = one-billionth of a metre, 0.000000001 m or 10^{-9} m.

Mass

Mt, Gt   A metric tonne (t) is 1,000 kilograms (kg). Carbon and carbon dioxide emissions are usually expressed in terms of millions (Mt; for megatonnes = 10^6 t) or billions (Gt; for gigatonnes = 10^9 t) of tonnes.

Time

ky	Thousands of years (= 1,000 or 10^3 years).
ka	Thousands of years before present ((= 1,000 or 10^3 years BP).
My	Millions of years (= 1,000,000 or 10^6 years).
Ma	Millions of years ago before present (= 1,000,000 or 10^6 years BP).

Volume

ppm(v)	parts per million (by volume); can also represented as 10^{-6} or 0.000001.
ML, GL	A litre is a fundamental measure of volume of water. One litre represents a 1 cubic decimetre (10x10x10 cm) of water, and weighs 1 kg. Engineering and environmental flows of water are generally calculated in millions (10^6; or ML, for megalitres) or billions (10^9; or GL, for gigalitres) of litres.

Acronyms

ACMA	Australian Communications and Media Authority
AGW	Anthropogenic (human-caused) global warming
AIMS	Australian Institute of Marine Science
AMO	Atlantic Meridional Oscillation
AO	Arctic Oscillation
ARGO	Integrated Ocean Observing System
BMO	British Meteorological Office (UK)
BOM	Bureau of Meteorology (Australia)
CERN	European Organisation for Nuclear Research
CETI	Central England Temperature Index (world's longest thermometer record)
CCN	Cloud condensation nuclei
CLIMAP	Climate: Long-range Investigation, Mapping and Prediction
COTS	Crown of thorns starfish
CRU	Climate Research Unit (branch of UK's Hadley Centre and MO)
CSIRO	Commonwealth Scientific and Industrial Research Organisation (Australia)
DAGW	Dangerous anthropogenic (human-caused) global warming
EEZ	Exclusive Economic Zone (marine territory)

El Niño	The warm condition of the ENSO climatic oscillation
ENSO	El Niño – Southern Oscillation
ETS	Emissions Trading Scheme
EU	European Union
FCCC	Framework Convention on Climate Change (a UN protocol)
GBR	Great Barrier Reef
GCM	General Circulation Model (deterministic computer model)
GeoNet	The New Zealand national natural hazard warning network
GNS Science	The former New Zealand Geological Survey
GISS	Goddard Institute of Space Studies (US, part of NASA)
GRASP	Geodetic Reference Antenna in Space (JPL NASA)
ICSU	International Council of Scientific Unions
IOD	Indian Ocean Dipole
IMO	International Meteorological Organisation (1873-1950)
IPART	Independent Pricing and Regulatory Tribunal (NSW)
IPCC	Intergovernmental Panel on Climate Change (a UN body)
ISSC	International Climate Science Coalition
JPL	Jet Propulsion Laboratory (NASA)
La Niña	The cold condition of the ENSO climatic oscillation
LIA	Little Ice Age
MRET	Mandatory Renewable Energy Tariff
MSU	Microwave sensing unit (satellite temperature measurement)
MWP	Medieval Warm Period
NAO	North Atlantic Oscillation
NASA	National Aeronautics and Space Agency (USA)
NIPCC	Nongovernmental International Panel on Climate Change
NIWA	National Institute of Water and Atmosphere (NZ)
OECD	Organisation for Economic Co-operation and Development
PDO	Pacific Decadal Oscillation
SOI	Southern Oscillation Index
TSI	Total Solar Insolation
UNEP	United Nations Environment Program
WMO	World Meteorological Association (1950-today)
WWF	World Wildlife Fund, or World Wide Fund for Nature

List of text tables

List of scientific figures

May 2011

Sources for figures

Fig. 1. British Meteorological Office, 2013. Met Office Hadley Centre observations datasets. CRUTEM4 Data. *http://www.metoffice.gov.uk/hadobs/crutem4/data/download.html.*Graph after Kahlbaum & Assoc. at Watts Up With That, 2012. CRU's new CRUTem4, hiding the decline yet again. *http://wattsupwiththat.com/2012/03/19/crus-new-hadcrut4-hiding-the-decline-yet-again-2/.*

Fig. 2. Mix, A.C. et al., 1995a. Benthic foraminiferal stable isotope record from Site 849, 0-5 Ma: Local and global climate changes. In: Pisias, N.G. et al. (eds.), Proceedings ODP, Scientific Results 138, pp. 371-412; and Mix, A.C., Le, J. & Shackleton, N.J., 1995b. Benthic foraminifer stable isotope stratigraphy of Site 846: 0-1.8 Ma. In: Pisias, N.G. et al. (eds.), Proceedings ODP, Scientific Results 138, pp. 839-56.

Fig. 3. Geerts, B. & Linacre, E. 1997. The height of the tropopause. *http://www-das.uwyo.edu/~geerts/cwx/notes/chap01/tropo.html.*

Fig. 4. Rohde, R. A., Global Warming Art. *http://en.wikipedia.org/wiki/File:CLIMAP.jpg.*

Fig. 5. Alley, R.B., 2004. GISP2 Ice Core Temperature and Accumulation Data, NOAA. Alley, R.B., 2000. The Younger Dryas cold interval as viewed from central Greenland. Quaternary Science Reviews 19, 213-226.

Fig. 6. Data after Alley, R.B., 2000. The Younger Dryas cold interval as viewed from central Greenland. Quaternary Science Reviews 19, 213-226. Interpretation after Davis, J.C. & Bohling, G.C., 2001. The search for pattern in ice-core temperature curves. American Association of Petroleum Geologists, Studies in Geology 47, 213-229.

Fig. 7. Temperature curve: Intergovernmental Panel on Climate Change (IPCC) 2001. Climate Change 2001: The Scientific Basis. Contribution of Working Group I to the Third Assessment Report of the IPCC. Ed.: J.T. Houghton, et al., Cambridge University Press. Synthesis Report, Fig. 2.3. Carbon dioxide curves after: Carbon Dioxide Information Analysis Center (CDIAC), 2010. *http://cdiac.ornl.gov/ftp/ndp030/global.1751_2006.ems.*

Fig. 8. Past Global Changes (PAGES) brochure, after Intergovernmental Panel on Climate Change (IPCC) 2001. Climate Change 2001: The Scientific Basis. Contribution of Working Group I to the Third Assessment Report of the IPCC. Ed.: J.T. Houghton, et al., Cambridge University Press.

Fig. 9. (above) Met Office Hadley Centre, 2013. HadAT: globally gridded radiosonde temperature anomalies from 1958 to present. *http://www.metoffice.gov.uk/hadobs/hadat/images/update_images/timeseries.png.* Thorne, P.W., et al., 2005. Revisiting radiosonde upper air temperatures from 1958 to 2002. Journal of Geophysical Research. 110: D18105, doi:10.1029/2004JD005753. (below) Spencer, R., 2013 (March 4). UAH Global Temperature Update for February, 2013: +0.18 deg. C. *http://www.drroyspencer.com/.*

Fig. 10. CETI: A study of Climatic Variability from 1660-2009. Willmore, N., 2009. A Detailed look at Europe. *http://climatereason.com/LittleIceAgeThermometers/CentralEngland_UK.html*

Fig. 11. (above) After an original figure in New Scientist, June 17, 1989. (below) Rohde, R., Global Warming Art. *http://commons.wikimedia.org/wiki/File:Milankovitch_Variations_large.png.*

Fig. 12. Humlum, O., 2013. Monthly Antarctic, Arctic and global sea ice extent since November 1978, after National Snow and Ice Data Center. *http://www.climate4you.com/.*

Fig. 13. Fisher, D., et al., 2006. Natural Variability of Arctic sea ice over the Holocene. EOS (Transactions of the American Geophysical Union) 87 (28), 273, 275, doi:10.1029/2006EO280001.

Fig. 14. Maue, R.N., 2013. Seasonal Tropical Cyclone Activity Update, Florida State University. *http://www.coaps.fsu.edu/~maue/tropical/.*

Fig. 15. Stephens, G.L. at al., 2012. An update on Earth's energy balance in light of the latest global observations. Nature Geoscience, (September). DOI: 10.1038/NGEO1580.

Fig. 16. Calculated graph from MODTRAN atmospheric model, University of Chicago. *http://modtran5.com/.*

Fig. 17. Spencer, R, 2008 (Oct. 20). Global Warming as a Natural Response to Cloud Changes Associated with the Pacific Decadal Oscillation (PDO). *http://www.drroyspencer.com/research-articles/global-warming-as-a-natural-response/.*

Fig. 18. Parrenin, F. et al., 2013. Synchronous change of atmospheric CO_2 and Antarctic temperature during the last deglacial warming. Science 339, 1060. DOi 10.1126/science.1226368.

Fig. 19. Rohde, R. A., 2013. Phanerozoic Carbon Dioxide, Global Warming Art. *http://www.globalwarmingart.com/wiki/File:Phanerozoic_Carbon_Dioxide_png.* Royer, D.L., et al., 2004. CO_2 as a primary driver of Phanerozoic climate. GSA Today 14 (3), 4-10. Fig. 2. Berner, R.A. & Kothavala, Z., 2001. GEOCARB III: A revised model of atmospheric CO_2 over Phanerozoic time. American Journal of Science 301, 182–204.

Fig. 20. Cubasch, U. & Weubbles, D. et al., 2012. IPCC WG1 Fifth Assessment

Report (Second Order DRAFT), Chapter 1: Introduction, Fig. 1.4. *http://wattsupwiththat.com/2012/12/14/the-real-ipcc-ar5-draft-bombshell-plus-a-poll/*.

Fig. 21. Scafetta, N., 2011. 'Testing an astronomically based decadal-scale empirical harmonic climate model versus the IPCC (2007) general circulation climate models' Journal of Atmospheric and Solar-Terrestrial Physics 80, 124-137. *http://www.sciencedirect.com/science/article/pii/S1364682611003385*.

Fig. 22. Douglass et al., 2007. A comparison of tropical temperature trends with model predictions. International Journal of Climatology DOI: 10.1002/joc.1651.

Fig. 23. (left) Spencer, 2013 (March 4). Global Microwave Sea Surface Temperature Update for Feb. 2013. *http://www.drroyspencer.com/2013/03/global-microwave-sea-surface-temperature-update-for-feb-2013-0-01-deg-c/*. (right) Argo, 2013. *http://www.argo.ucsd.edu/*; Evans, D., 2011. *http://joannenova.com.au/2011/12/the-travesty-of-the-missing-heat-deep-ocean-or-outer-space/#more-19151*.

Fig. 24. Australian Bureau of Meteorology, National Tidal Centre, 2009. Mean Sea Level Survey 2009, 11 pp., Figs. 1 and 2.

Fig. 25. Hannah, J. & Bell, R.G., 2012. Regional sea level trends in New Zealand. Journal of Geophysical Research: Oceans 117, DOI: 10.1029/2011JC007591. Inset: Watson, P.J., 2011. Is there evidence yet of acceleration in mean sea level rise around mainland Australia? Journal of Coastal Research 27, 368-377.

Fig. 26. (left) Liu, Y. et al., 2009. Instability of seawater pH in the South China Sea during the mid-late Holocene: Evidence from boron isotopic composition of corals. Geochimica et Cosmochimica Acta 73: 1264-1272. (right) Segalstad, 2013. *http://folk.uio.no/tomvs/esef/esef0.htm*. Bethke, C.M., 1996. Geochemical Reaction Modeling. Oxford University Press, New York, Fig. 6.1 (p. 84). Skirrow, G., 1965: The dissolved gases – carbon dioxide. In: Riley, J.H. & Skirrow, G. (Eds.): Chemical Oceanography. Academic Press, London, pp. 227-322.

Fig. 27. (above) Svensmark & Friis-Christensen, E., 2007. Reply to Lockwood and Frohlich – The persistent role of the Sun in climate forcing. Danish National Space Centre, Scientific Report 3/2007, 6 pp. (below) Neff, U., et al., 2001. Strong coherence between solar variability and the monsoon in Oman between 9 and 6 kyr ago. Nature 411, 290-293.

Fig. 28. Figure courtesy Dr Willie Soon (Smithsonian Centre for Astrophysics): Soon, W. & Legates, D.R., 2013. Solar irradiance modulation of Equator-to-Pole (Arctic) temperature gradients: Empirical evidence for climate variation on multi-decadal timescales. Journal of Atmospheric and

Solar-Terrestrial Physics 93, 45–56. Bekryaev, R., Polyakov, I. & Alexeev, V., 2010. Role of polar amplification in long-term surface air temperature variations and modern Arctic warming. Journal of Climate 23, 3888-3906. Muller, R.A. et al., 2012. Earth Atmospheric Land Surface Temperature and Station Quality in the United States. JGR Special Publication, The Third Santa Fe Conference on Global and Regional Climate Change, manuscript number 12JD018146. Zhou T. & Yu, R, 2006. Twentieth-Century Surface Air Temperature over China and the Globe Simulated by Coupled climate Models. Journal of Climate 19, 5843-5858.

Fig. 29. Skinner, B.J., Porter, S.C. & Botkin, D.B. 1999 (2nd ed.). *The Blue Planet: An Introduction to Earth System Science.* John Wiley & Sons, New York, Fig. 11.15.

Fig. 30. Figure courtesy Chris de Freitas and John McLean. De Freitas, C.R. & McLean, J.D. 2013. Update of the chronology of natural signals in the near-surface mean global temperature record and the Southern Oscillation Index. International Journal of Geosciences 4, 234-239.

Fig. 31. JISAO, 2013. The Pacific Decadal Oscillation (PDO), *http://jisao. washington.edu/pdo/.* Giorgiogp2, 2013. Monthly values for the Pacific decadal oscillation index, 1900-2012. *http://en.wikipedia.org/wiki/File:PDO.svg.*

Fig. 32. McCulloch, M. et al., 2003, Coral record of increased sediment flux to the inner Great Barrier Reef since European settlement. Nature 421, 727-730.

Fig. 33. Australian Bureau of Meteorology (BOM), 2013. Australian climate variability and change – Time series graphs. *http://www.bom.gov.au/cgi. bin/ climate/change/timeseries.cgi?graph=rain&area=aus&season=0112&ave_yr=5.*

Fig. 34. Australian Bureau of Meteorology (BOM), 2011. Known floods in the Brisbane & Bremer River Basin. *http://www.bom.gov.au/qld/flood/ fld_history/brisbane_history.shtml.*

Figs. 35. Marohasy, J. 2003. Myth & the Murray. Measuring the real state of the river environment. IPa Backgrounder, December 2003, Fig. 1. *http://aefweb.info/data/Myth%20&%20the%20Murray.pdf.* Murray-Darling basin commission, 2005. The Living Murray, Foundation Report on the significant ecological assets targeted in the First Step Decision, Fig. 7.5. MDBC, Canberra.

Fig. 36. CSIRO National Flagship, Climate Adaptation. Presentation at Marcus Odhum College, August 2008. Data from Murray-Darling Basin Commission (2008).

Fig. 37. Australian Bureau of Meteorology (BOM), 2013. Tropical Cyclone Trends. *http://www.bom.gov.au/cyclone/climatology/trends.shtml.*

Fig. 38. Nott, J., Haig, J., Neil, H. & Gillieson, D., 2007. Greater frequency variability of landfalling tropical cyclones at centennial compared to seasonal and decadal scales. Earth & Planetary Science Letters 255, 367-3.

Fig. 39. McLean, J., 2011. Sea-surface Temperatures along the Great Barrier Reef. *http://mclean.ch/climate/GBR_sea_temperature.htm.*

Fig. 40. Muller, R.A., 2012. Naked Copenhagen: The Numbers Behind the OpEd. *http://mullerandassociates.com/resources/naked-copenhagen/.*

Fig. 41. Australian Greenhouse Office, 2002. National Greenhouse Gas Inventory.

Fig. 42. Akosofu, S.-I., 2009. Two Natural Components of the Recent Climate Change. *http://people.iarc.uaf.edu/~sakasofu/pdf/two_natural_components_recent_climate_change.pdf.*

Fig. 43. Abdussamatov, H.I., 2012. Bicentennial decrease of the Total Solar Irradiance leads to unbalanced thermal budget of the Earth and the Little Ice Age. Applied Physics Research 4. *www.ccsenet.org/apr.*

November 2011

Recommended reading and reference material

The research papers that underpin many of the scientific statements made in this book were not written with the layperson in mind. Therefore, instead of references to these papers, we provide here a short list of recommended additional books and websites.

Sources listed in Sections B-D have been selected mainly for their readability and accessibility. Nonetheless, and as for the IPCC and NIPCC reports listed in Section A, many of those items in B-D also contain copious references to the original, peer-reviewed scientific literature.

A. Major compilations of evidence for and against the occurrence of dangerous man-made global warming.

Intergovernmental Panel on Climate Change, 2009. *Fourth Assessment Report. Working Group I Report. The Physical Science Basis* (free download or order from *http://www.ipcc.ch/ipccreports /ar4-wg1.htm*).

Highly technical. The official source of advice on climate change to all governments, worldwide. Summarises much excellent science, but the more alarmist conclusions are based on many hidden assumptions, and have been heavily challenged by independent scientists.

Nongovernmental International Panel on Climate Change. Singer, S.F. & Idso, C., 2009. *Climate Change Reconsidered*, 880 pp. *http://www.nipccreport.org/*.

The due diligence counterpart to the IPCC reports. Somewhat technical, but a comprehensive and independent critical assessment that includes summaries of many scientific papers that are not taken into account in IPCC reports, and which fail to find evidence for dangerous human-caused global warming.

B. Introductory and background material on climate change.

Burroughs, W. (ed.), 2003. *Climate into the 21st Century.* World Meteorological Organisation & Cambridge University Press, 240 pp.

A well-illustrated, well-organised and generally well-balanced introduction to the major elements of meteorology and climate change.

Kininmonth, W., 2004. *Climate Change: a Natural Hazard.* Multi-Science Publishing Company, Brentwood, U.K.

Somewhat technical, but an interesting and authoritative summary of the basic meteorology and physical principles that determine climate processes.

Ruddiman, W.F., 2001. *Earth's Climate, Past & Future.* Freeman & Company, New York, 465 pp.

A comprehensive introductory text which covers climate change well and across the board. Contains careful, accurate, well-illustrated and well-balanced explanations of most climate topics.

C. Other readable books, several of which encompass the middle-ground view that both natural and possible human-caused climate hazard should be planned for the same way – by preparation and adaptation

Alexander, R.B., 2012. *Global Warming False Alarm – The Bad Science behind the United Nations' assertion that Man-Made CO_2 causes Global Warming.* Canterbury Publishing, Michigan, 200 pp.

An easy-to-read and sober assessment of the poor quality of much of the science that is cited in support of dangerous warming, written by an experienced environmental scientist with a research degree in Physics.

Booker, C., 2009. *The Real Global Warming Disaster. Is The Obsession With 'Climate Change' Turning Out To Be The Most Costly Scientific Blunder In History?* Continuum, 368 pp.

The best and most detailed account of the history and socio-political pathology of the global warming issue.

Carter, R.M., 2010. *Climate: the Counter Consensus.* Stacey International, London, 315pp.

An easy-to-read and comprehensive summary of both the science and politics of the global warming issue.

Laframboise, D., 2011. *The Delinquent Teenager who was Taken for the World's Top Climate Expert.* Connorcourt Publishing. *http://nofrakkingconsensus.com/my-book/.*

A searing expose of the scientific and political malfeasance associated with the IPCC.

Essex, C. & McKitrick, R., 2007 (2nd ed.). *Taken by Storm. The Troubled Science, Policy and Politics of Global Warming.* Key Porter paperback (available from Amazon CANADA).

Delightfully written, insightful and whimsical account of many of the key issues of global warming science.

Etherington, J., 2009. *The Wind Farm Scam.* Stacey International, Independent Thinkers series.

A clearly-stated and excellent account of the deficiencies of wind-power generation.

Gerhard, L.C. et al., 2001. *Geological Perspectives of Global Climate Change.* American Association of Petroleum Geologists, Studies in Geology #47 (available from AAPG website).

Technical, but all the papers in the volume are clearly written, well illustrated and easy to read. Perhaps the best single collection of papers on a wide range of major climate change issues.

Lawson, N., 2007. *An Appeal to Reason: A Cool Look at Global Warming.* Duckworth Overlook, 149 pp.

A short and very well written account, with particular reference to the economic and social aspects of global warming hysteria.

Michaels, P.J. (ed.), 2005. *Shattered Consensus: The True State of Global Warming.* Rowman and Littlefield Publishers, Oxford, 292 pp.

A good introduction to the problems and pitfalls of the global warming debate, with individual chapters written by acknowledged experts in each field.

Montford, A., 2010. *The Hockey Stick Illusion: Climategate and the Corruption of Science.* Stacey International, Independent Thinkers series.

The definitive account of the Mann, Bradley & Hughes hockey stick saga.

Nova, J., 2009. *The Skeptics Handbook.* Free download or order from: *http:// joannenova.com.au/global-warming/.*

An attractive, witty and succinct account of the main points in the global warming debate, written in a lively fashion that is suitable for secondary school pupils as well as adults.

Paltridge, G.W., 2009. *The Climate Caper.* Connor Court, 111 pp.

A brief, clearly written and fascinating account of the science and sociology of the global warming phenomenon with especial respect to Australia, written by a former CSIRO Chief Research Scientist.

Plimer, I., 2009. *Heaven and Earth: Global Warming – The Missing Science.* Connor Court, 503 pp.

The geological viewpoint. A profusely referenced account of the history of natural climate change seen through the eyes of a senior research scientist who has long been a doughty warrior for scientific truth.

Spencer, R., 2008. *Climate Confusion. How Global Warming Hysteria Leads to Bad Science, Pandering Politicians and Misguided Policies That Hurt the Poor.* Encounter Books, N.Y., 191 pp.

The meteorological viewpoint. A very readable recounting of important facts about climate change processes, and their implications.

D. Independent climate change websites with information on global warming and climate change

This brief list does not include the major public database and reference sites on climate such as NASA, NOAA, CRU, BOM, NIWA and other national meteorological offices. These are widely known and easily found through web search engines. Listed here instead are some selected and less obvious, but independent and critical, sources of information.

D'Aleo, J. – IceCap – *http://icecap.us/index.php.*
Humlum, O. – Climate4you – *http://www.climate4you.com/.*
Idso, C. & S. – CO 2 Science – *http://www.co2science.org/.*
International Climate Science Coalition – *http://www.internationalcsc.org.au/.*
McIntyre, S. – Climate Audit – *http://climateaudit.org/.*
Morano, M. – Climate Depot - *http://www.climatedepot.com/*
Nova, J. – JoNova – *http://joannenova.com.au/.*
Science and Public Policy Institute (SPPI) – *http://scienceandpublicpolicy.org/.*
Watts, A. – Watts Up With That – *http://wattsupwiththat.com/*

Finally, Quadrant Magazine has published a number of critiques and rebuttals of recent official reports by the Australian Department of Climate Change, the Climate Commission and CSIRO, which can be accessed under the heading Global Warming: an essential reference, at the following website: *http://www.quadrant.org.au/blogs/doomed-planet/2011/04/due-diligence-reports.*

The Authors

Robert (Bob) Carter, B.Sc. Hons (Otago, NZ), Ph.D. (Cambridge, UK) is a marine geologist and environmental scientist with more than 40 years professional experience who has held academic staff positions at the University of Otago (Dunedin), the University of Adelaide (Adelaide) and James Cook University (Townsville), where he was Professor and Head of School of Earth Sciences between 1981 and 1999. His career has included periods as a Commonwealth Scholar (Cambridge University), a Nuffield Fellow (Oxford University) and an Australian Research Council Special Investigator. Bob Carter's current research on climate change, sea-level change and stratigraphy is based on field studies of Cenozoic sediments (last 65 million years) from the Southwest Pacific region and the Great Barrier Reef. Carter has served as Chair of the Earth Sciences Discipline Panel of the Australian Research Council, Chair of the Australian Marine Science and Technologies Committee, and Director of the Australian Office of the Ocean Drilling Program. Bob has acted as an expert witness on climate change before the U.S. Senate Committee of Environment & Public Works, the Australian and N.Z. parliamentary Select Committees into emissions trading, and was a primary science witness in the U.K. High Court case of Dimmock v. H.M.'s Secretary of State for Education, the 2007 judgement from which identified nine major scientific errors in Mr Al Gore's film *An Inconvenient Truth*. Carter is author of the book, *Climate: the Counter Consensus* (2010, Stacey International Ltd., London), and is currently an Emeritus Fellow of the Institute of Public Affairs.

Martin Feil (Arts-Law degree) is an economist specialising in Customs, logistics, ACCC actions, industry policy and international trade related matters, including transfer pricing. Feil got his first job in the Customs department, and then became the Industries Commission's youngest project director at the age of 26. He was eventually responsible for 11 major industry inquiries, before striking out on his own and working as an industry-policy consultant for the next 30 years. During that time he also owned trucks, warehouses, Customs bonds-stores, and container yards, and worked for the Australian Taxation Office as one of the few Australian independent experts on transfer pricing and profit repatriation by multinationals. Feil has been chairman of the Institute of Chartered Accountants' Customs committee, and the institute's representative on the tax office's transfer-pricing subcommittee. Feil is the author of the book, *The Failure of Free-Market Economics* which was, fittingly, illustrated by John Spooner.

Stewart W. Franks, B.Sc. Hons, Ph.D. (Lancaster, UK) is a hydro-climatologist who is the foundation Professor of Environmental Engineering at the University of Tasmania, Hobart. Stewart's research interests centre on the quantification and reduction of uncertainty in environmental modelling and hydro-climatic risk assessment, including modelling land surface – atmosphere interactions for atmospheric/climate models. He has published extensively on the role of ENSO, IOD and PDO on Australian flood, drought and bushfire risk. He is currently President of the International Commission on the Coupled Land-Atmosphere System (ICCLAS), a commission of the International Association of Hydrological Sciences (IAHS). The remit of the commission is to organise symposia and workshops on dealing with hydrological variability and the interactions between the land surface and the atmosphere. A special focus is directed toward building knowledge and capacity in developing countries. Stewart has edited a number of books documenting examples of historic hydro-climatic variability across the globe.

William Kininmonth, B.Sc. (UWA), M.Sc. (Colorado State, USA), M.Admin. (Monash), is a consulting climatologist with more than 45 years professional experience. He worked with the Australian Bureau of Meteorology for 38 years in weather forecasting, research and applied studies; for 12 years until 1998 Bill was head of its National Climate Centre. He has worked closely with the World Meteorological Organisation since 1982 as Australia's delegate to the Commission for Climatology, in expert working groups, lecturing at regional training seminars, and later as a consultant. He has been a member of Australian delegations to international conferences and intergovernmental negotiations relating to climate, including for the UN's Framework Convention on Climate Change (1991-92). William Kininmonth participated in the Australian Public Service Executive Development Scheme (1977) and was leader of an Australian Government project of assistance to the Meteorology and Environmental Protection Administration of the Kingdom of Saudi Arabia (1982-85). Kininmonth is author of the book, *Climate Change: A Natural Hazard* (2004, Multi-Science Publishing Co, UK).

Bryan Leyland, M.Sc., FIEE(rtd), F.I.Mech.E., FIPENZ, is a consulting engineer with an international background in hydropower and new renewable energy technologies, including wind, solar, tidal, and wave power. He has expertise in combined cycle power generation, diesel generation and nuclear power, and wide experience in electricity transmission and distribution. Leyland has provided consulting services to the Asian Development Bank, the World Bank, a large water storage project in India and a hydro development in Mongolia. He has worked in many developing countries including Bhutan, the Pacific Islands, Bangladesh, Sierra Leone, Niger and Mauritius. He has presented many papers at international conferences on hydropower and was named by the international journal *Waterpower and Dam Construction* as one of the 60 most influential people in the hydro business worldwide. Leyland has been interested in the debate on man-made global warming for many years and has contributed to climate conferences in Chicago, New York and Stockholm. He has written a number of articles pointing out the serious economic problems surrounding new energy technologies, and highlighting the fact that they are an ineffective and very expensive way of reducing carbon dioxide emissions. He has made many public presentations pointing out that the balance of evidence is strongly against the hypothesis that man-made global warming is real and dangerous. As Leyland and his wife are majority owners of a small hydropower scheme that profits from New Zealand's Emissions Trading Scheme, his stance as a climate sceptic is against his own financial interests.

John Spooner, LLB (Monash) practised as a lawyer for three years before he commenced drawing for *The Age* newspaper in 1974, finally leaving the Law altogether in 1977. John has presented many exhibitions and won many awards for his prints and drawings, including the following:
1982: Walkley Award. 1984: Two man exhibition, prints and drawings, with fellow artist Peter Nicholson, Powell Street Gallery, Melbourne. 1982: Spooner—*Drawings Caricatures and Prints*, published by Thomas Nelson. In the period 1985/1986 Spooner was the winner of five Stanley Awards, including the Black and White Artist of the Year Gold Stanley Award. 1986: Co-winner Mornington Peninsula Print Acquisitive Award. 1986: Co-winner of the Fremantle Print Prize. 1989, *Bodies and Souls*, a book of his drawings, caricatures and prints, published by Macmillan, forward by Barry Humphries. 1994 Spooner was awarded two Walkley Awards for Best Illustration and Best Cartoon. 1996: Solo exhibition John Spooner: A Survey, Prints Drawings and Paintings 1987-1996 was held at the Westpac Gallery. 2000: Solo exhibition,

Recent Thoughts at George Adams Gallery, (Victorian Arts Centre) showing paintings, cartoons and printmaking. 2000: Solo exhibition at McClelland Gallery, Victoria—Prints, Paintings and Cartoons from the Year 2000. 2003: Received the Graham Perkin Award for Australian journalist of the year, Quill Awards. 2004: Solo Exhibition — Etchings, Cartoons & Paintings at Chrysalis Publishing Gallery and Studio, Melbourne. 2007: Solo Exhibition — New Works at Chrysalis Gallery and Studio, Melbourne.

Spooner's works are represented in the collections of The National Gallery of Australia, National Library of Canberra, The National Gallery of Victoria, The Victorian State Library Art, Ballarat Gallery, The Melbourne Cricket Club Museum and in other public and private collections throughout Australia and internationally. His publications include the book *A Spooner in the Works*, published in 1999 by Text Publishing, comprising cartoons, prints and paintings.

June 2012

INDEX

A

Abdussamatov, Habibullo
  solar cycles control climate · 233
acidification of the ocean · *See* ocean:
changing alkalinity
ad hominem abuse · 55
Akosofu, Syun-Ichi
  temperature recovery from Little Ice Age
  · 232
albedo · 91, 104, 128
Alternative Energy · See XII
alternative energy
  biofuels · 220
  capacity factor · 212
  comparative costs · 213
  nuclear · 222
  solar power · 217
  tidal power · 218
  what's wrong with coal-fired power
  stations · 224
  windpower · 211
    does it save emissions · 216
    how environmentally friendly is it ·
215
Anderegg, William
  analysis of publications of climate
  scientists · 65
Anderson, Kevin
  Tyndall Centre, Manchester University ·
  165
Arctic Ocean sea ice · 7, 28, 82
Arctic Oscillation · 158
ARGO ocean observing system · 118
Australia
  drought · 8, 162
  fewer cyclones in 20th century · 173
  greatest natural hazards · 231
  historic temperature records · 160
  influence of geological setting on climate
  · 163
  influences on climate · 165
  is global warming threatening the GBR ·
  175 natural aridity · 163
  prone to floods · 165
Australian Climate · See VIII

Australian Communications & Media
Authority · 196
Australian Institute of Marine Science
  Great Barrier Reef research · 175
author biographies · 255

B

Banqiao dam collapse, China
  largest hydropower accident · 223
Bentek, Bryce
  Texas windfarms · 217
biodiesel, biofuels · *See* alternative energy,
biofuels
bioethanol · *See* alternative energy, biofuels
Blair, Tony
  commissioned Stern Report · 180
Bolt, Andrew · 7
  warming to be averted by emissions cuts
  · 202
Brabeck-Letmathe, Peter
  stupidity of using biofuels · 222
British Broadcasting Corporation
  negligent in reporting Climategate · 10,
  53
British Meteorological Office
  acknowledgement that warming has
  ceased · 131
Brown, Lester
  stupidity of using biofuels · 222
Brunner, Ronald and Amanda Lynch
  Adaptive Governance and Climate Change
  · 235
Bureau of Meteorology
  enhanced bushfire danger · 135
Burton, Justice Michael
  UK High Court case against Mr Gore's
  film · 46

C

carbon · *See* carbon dioxide
carbon dioxide
  absorbed in Australia's EEZ · 201
  are modern levels unusually high · 112
  beneficial for the biosphere · 110
  climate sensitivity · 103
  current level in the atmosphere · 112

Pliocene climatic optimum · 27, 32
polar bears
    are not decreasing · 85
precaution against natural hazard · 227
precautionary principle
    giving earth the benefit of the doubt · 228
Productivity Commission · 185
    quiet on the carbon dioxide tax · 206
proxy records · 27, 31
    of temperature · 72
oxygen isotope analysis · 173

R

radioactive decay
    heating from · 109
Reiter, Paul
    Director infectious diseases, Pasteur
    Institute · 4
Riehl, Herbert and Joanne Malkus
    cumulus clouds as a cooling mechanism
    · 99
Rio Earth Summit, 1992 · 67, 69
Rudd, Kevin
    commissioned Garnaut Report · 180

S

Santer, Ben
    signal-noise ratio of human climate signal
    · 59
Scafetta, Nicola
    solar cycles control climate · 122
sceptic · See climate sceptic
science court · 62
scientific consensus · 2, 68
scientific method · 2, 48, 110
sea-ice
    Arctic Ocean and Antarctica · 7, 82
sea-level · 133
    difference between local and global
    (eustatic) · 133
    higher than present · 28
    isostatic rebound · 135
    rate of rise lessening · 136
    satellite measurement · 136
    tide gauge measurement · 136
    what controls the position of the

shoreline · 138
Singer, Fred
    Director, U.S. Satellite Weather Service · 45
solar influences on climate · 150
    1,500 year (Bond) cycle common in
    Holocene · 81
    Grand Minima and Maxima · 152
    latitudinal distribution of radiation · 19
    magnetic activity · 148
    modulation of cosmic rays · 152
solar power · See alternative energy
Soon, Willie
    solar influences on climate · 152
Southern Oscillation Index · 86, 143, 153
speleothems, Chillagoe cave · 173
Steffen, Will
    Director, Climate Change Institute, ANU
    · 50
Steirou and Houtsoyiannis
    reanalysis, 20th century temperatures · 33
Stern, Nicholas
    author of UK report on the economics of
    global warming · 180

T

tectonics · 231
temperature
    150 year thermometer record represents
    weather not climate · 30
    is not an intrinsic property · 25
temperature trend
    long-term cooling through the Holocene
    · 87
The Carbon Dioxide Tax – I · *See* **IX**
The Carbon Dioxide Tax – II · *See* **X**
The Carbon Dioxide Tax – III · *See* **XI**
The Greenhouse Hypothesis · *See* **IV**
troposphere
    definition of · 127
Tyndall, John
    first measurements of radiation absorbtion
    by greenhouse gases, 1850s · 97

U

United Nations
    letter to Secretary-General (Bali, 2007)
    · 67